ASIAN STUDIES ASSOCIATION OF AUSTRALIA

Southeast Asia Publications Series

I0643870

MADURESE SEAFARERS

Prahus, Timber and Illegality on the Margins of the Indonesian State

ASIAN STUDIES ASSOCIATION OF AUSTRALIA
Southeast Asia Publications Series

Since 1979 the Southeast Asia Publications Series (SEAPS) has brought some of the best of Australian scholarship on Southeast Asia to an international readership. It seeks to publish leading-edge research by both young and established scholars on the countries and peoples of Southeast Asia across all disciplines of the humanities and social sciences with particular encouragement to interdisciplinary and comparative research.

SEAPS is now published for the Asian Studies Association of Australia by NUS Press, a unit of the National University of Singapore, in alliance with the University of Hawaii Press in North America and the Nordic Institute of Asian Studies (NIAS) [non-Indonesia titles] and the KITLV Press [Indonesia titles] in Europe.

Website: http://iceaps.anu.edu.au/asaa_publications/southeast_asia.html

MADURESE SEAFARERS

Prahus, Timber and Illegality on the Margins of the Indonesian State

Kurt Stenross

Asian Studies Association of Australia
in association with
UNIVERSITY OF HAWAI'I PRESS
HONOLULU

Published in Singapore by:
NUS Press
National University of Singapore
AS3-01-02, 3 Arts Link
Singapore 117569

Published in North America by:
University of Hawai`i Press
2840 Kolowalu Street
Honolulu, HI 96822 USA
www.uhpress.hawaii.edu

Library of Congress Cataloging-in-Publication Data

Stenross, Kurt, 1946–
Madurese seafarers: prahus, timber and illegality on the margins of the Indonesian state / Kurt Stenross.
p. cm. – (Southeast Asia publication series)
Includes bibliographical references and index.
ISBN 978-0-8248-3555-2 (pbk.: alk. paper)

1. Madurese (Indonesian people) – Commerce. 2. Shipping – Indonesia. 3. Praus – Indonesia. 4. Indonesia – Commerce – Government policy. 5. Lumber trade – Government policy – Indonesia. I. Title. II. Series: Southeast Asia publications series.

DS632.M23.S84 2011
305.89'92234 — dc22

2010034371

Cover: A traditional Madurese sailing transport prahu, a *janggolan*, carrying a cargo of poles from the island of Beliton, in the Java Sea, to Jakarta, 1984. The hold is filled to the deckhouse roof.

Printed in Singapore

Contents

Lete-lete from Sumenep with a cargo of timber from South Sumatra, bound for Jakarta, 1984

List of Maps

List of Photographs

All the photographs in this book were taken by the author and are from his personal collection.

List of Figures

Conventions

Ethnicities. Although I would prefer to avoid the use of the suffix '-ese' because of its orientalist connotations, I have retained it where it is necessary to distinguish an indigenous people from the name of their place of origin. Thus while this work refers to groups of 'Madurese', 'Javanese', 'Makassarese' and 'Butonese' ethnicity, it also refers to groups of 'Bugis' and 'Mandar' ethnicity.

Names of indigenous vessel types. Names of indigenous vessel types have been italicised throughout. No plurals are given for these words, with plurality being apparent by context. The exception is the word 'prahu', which is an anglicised version of the Indonesian word *perahu*, and is sufficiently widely known to be treated as a loan word in English. Both 'prahu' and the alternative borrowed form, 'prau', have long been used without italics in English language works, including British pilot volumes.

Names of timber species. Indigenous names of timber species have been italicised in the first instance. While readers may be unfamiliar with these names, some of them are used internationally for trade purposes and have appeared in numerous publications without italicisation.

Ordinary Indonesian words. These have been italicised in the first instance, except for abbreviations or names of organisations. Again, no plural form is used. For readers without any knowledge of Indonesian, [c] is always pronounced [ch], as in 'church', [g] is always hard, as in 'get', [j] is pronounced as in 'jump', and [y] is pronounced as in 'young'. For some types of vessel I have indicated the proper pronunciation, because experience has shown that these words are otherwise likely to be pronounced incorrectly.

Glossary

afdeeling	administrative district under Dutch rule
Airud	Polisi~, water (and air) police (*air*, 'water'; *udara*, 'air')
alis-alis (*lis-alis*)	type of small traditional open vessel of Madura
bangkirai	kind of tree (*shorea laevifolia Endert*)
becak	pedal-driven trishaw
beluntu	vessel from Bawean with hull form of European style, but with traditional sailing rig and lateral rudder
benuas	kind of tree (*shorea laevis Ridl.*)
bifid-ended	(of boats) having a stem or sternpost with a bifurcated appearance when viewed in profile
blater	forceful, tough (of person) (Java, Madura)
bungur	kind of tree (*lagerstroemia* sp.)
bupati	the highest-ranking official of a *kabupaten*
cemplon	type of traditional decked transport vessel of Java
desa	local administrative unit or ward, lowest tier of the formal government structure
Dinas Kehutanan	Forestry Service
DPP	*daftar penggantian pengangkutan*, change of mode of transportation (permit document)
DR	*dana reaboisasi*, reforestation fund levy
golekan	type of traditional transport vessel of Madura
Haji	title for a Muslim who has made the pilgrimage to Mecca
HPH	*hak pengusahaan hutan*, forestry concession permit
jago	a 'tough', person prepared to use force for coercion
janggolan	traditional type of transport vessel of Madura
jong	traditional Southeast Asian ship, extant until the seventeenth century
kabupaten	sub-provincial administrative district

kacik	type of traditional transport vessel of Madura, based on the *cemplon* of Java
kampung	small village or hamlet
kapur	various related kinds of tree (*dryobalanops* sp.)
kecamatan	administrative sub-district, below the level of a *kabupaten*
kelotok	small canoe-like open motor boat, used in Kalimantan
kelurahan	village-level administrative unit, similar to a *desa*
kepala desa	elected official in charge of a *desa* (*kepala*, 'head')
keruing	various related kinds of tree (*dipterocarpus* sp.)
ketupat	rice cake wrapped in plaited coconut leaves; festival with giving of such cakes
kiai (*kiyayi*)	charismatic religious leader (Java, Madura)
klebun	Madurese term for *kepala desa*
KL	*kapal layar*, sailing vessel
KLM	*kapal layar motor*, motor-sailing vessel
KM	*kapal motor*, motor vessel
kolek	any open boat (Java); a particular type of *mayang* fishing vessel of Central or West Java
konteng	large open fishing vessel of *mayang* family
koyan	28–30 *pikul* or more depending on the commodity
KPM	Koninklijke Paketvaart Maatschappij (Royal Packet Navigation Company)
Kuala	(in place names) estuary, river
laban	kind of tree (*vitex* sp.)
lambo	non-traditional vessel, especially with counter stern in European style, widely used in eastern Indonesia
last	4,000 pounds; nominal or 'manifest' ton of cargo (Dutch)
lete-lete	type of traditional transport vessel of East Madura
mayang	various types of closely related vessels used in traditional net fishery
MC	*meranti campur*, mixed meranti species of timber
meranti	various related kinds of tree (*shorea* sp.)
MSM	Madoera Stoomtram Maatschappij (Madura Steamtram Company)
nyamplong	kind of tree (*calophyllum inophyllum*)
padewakang	type of large decked trading vessel of Sulawesi, with two masts and bowsprit (nineteenth century)

pajak laut	'sea tax' (unauthorised payment demanded by water police)
pajala	basic vessel type of South Sulawesi, often undecked; also a small decked vessel similar to the *padewakang*
Pak (Bapak)	Mr, term of respect for older man or official; father
palari	type of trading vessel from Sulawesi, developed from the *padewakang* (late nineteenth–twentieth centuries)
peduang (*paduwang*)	traditional double-outrigger vessel of Madura
Pelayaran Nusantara	Inter-island shipping (official category for modern ships)
Pelayaran Rakyat	People's Shipping (official category for prahu shipping)
Pelra	association of *Pelayaran Rakyat* entrepreneurs
pencalang	type of traditional Javanese trading vessel of Malay origin (seventeenth–nineteenth centuries)
percaton	appanage system in Madura existing until the late nineteenth century
pesantren	residential Islamic learning centre
pikul	about 125 pounds (load of two baskets carried on a pole)
pinis (*pinisi*)	type of large trading vessel used by Bugis and Makassarese, with European sailing rig (twentieth century)
prahu	any wooden vessel, especially one used for transport work
PSDH	*provisi sumber daya hutan*, provision for forest resources levy
PT	*perusahaan terbatas*, limited liability company
rakyat	common people
rato	ruler, prince
retribusi	(official) levy
SAKO	*surat angkatan kayu olahan*, timber transport permit
sampan	general term for small open vessels on the south coast of Madura, with no similarity in form to Chinese sampan
sawah	wet-field rice farming, plot of land for same

sekoci	non-traditional prahu after European style, with sharp stern and central rudder; also very small flat-bottomed craft
sepel	non-traditional type of prahu, but with lateral rudder
SKSHH	*surat keterangan sahnya hasil hutan*, certificate of legality of forest produce
sloop	~ rig, a sailing rig with one mast and two sails (mainsail and jib)
sope (*sopek*, *sopet*)	various types of small single-masted traditional vessels of South Sulawesi origin
suku	ethnic or sub-ethnic group
Sungai	(in place names) river
Tanjung	(in place names) point, cape, promontory
tegal	dry-field farming
toop	large nineteenth-century prahu with two or three masts, similar in hull form to European vessels but often with lateral rudders
transom	(of vessels) transverse panel-like structure at one or both ends of a hull
ulin	kind of tree (*eusideroxylon zwageri*)
VOC	Vereenigde Oost-Indische Compagnie, Dutch East India Company
warung	small temporary or mobile food stall

Preface

My interest in Indonesian watercraft and maritime culture was aroused during a two-month visit to Indonesia in 1974. On subsequent visits over the next few years I experienced more of this apparently anachronistic maritime world, mainly in eastern Indonesia. Yet, notwithstanding this varied experience, on my first visit to Madura in 1983 I was amazed by the vessels I saw, which were like no others in the archipelago. I resolved then to learn as much as I could about these unique craft while it was still possible to do so, before they succumbed to the trend of modernisation already taking place elsewhere among the indigenous transport fleets of Indonesia. Throughout the rest of the 1980s I returned each year during my holidays to Java and Madura, visiting every maritime village I could and even taking passages around the Madura Strait and across the Java Sea on traditional vessels.

Yet as my knowledge deepened of the technical and practical aspects of these vessels, I became increasingly aware of how little I really knew about the maritime Madurese themselves. I had only a vague understanding of Indonesian history; I had no idea as to why the Madurese had become a great maritime people while their near neighbours, the Javanese and Balinese, were evidently not; nor did I have any idea of why the boatbuilding traditions of the Madurese were so different from those of South Sulawesi and eastern Indonesia.

With my interest in history kindled, I decided to enter university, majoring in Southeast Asian Studies. As part of my undergraduate course, in 1997–98 I spent a year in Indonesia. During the second semester there I undertook a three-month field project, staying in a prahu village in Madura. This field study deepened into my Honours year research project, about the *janggolan*, a type of traditional Madurese vessel, and was followed by further visits to Madura in 1999 and 2000.

By this time I was well aware of the paucity of literature about the seafaring Madurese. I was also well aware that the traditional maritime world, so recently flourishing, had all but disappeared as a result of the rapid changes that had taken place from the mid-1980s onwards. In this situation, I felt uniquely placed to undertake a social historical study of the maritime Madurese in the spirit proposed by Macknight (1980), who had noted the dearth of anthropological studies in the literature concerning prahus and urged researchers to move beyond narrowly based descriptions of the vessels themselves and focus as well on the social, cultural and economic contexts in which they were operated.

I had a few advantages in undertaking such an ambitious study. First, I had the great advantage of having observed this group intermittently over a prolonged period of economic change, with plenty of participation as well as observation. Second, I had a deep interest in the traditional vessels and had taken much trouble to understand the details of their construction, sailing rigs and operating techniques, and the ornamental and symbolic elements that were of such importance in this culture. Although this study is not directly concerned with these matters, my interest in them led to a deepening of my interest in Madurese maritime society and culture and its place in the wider maritime culture of Southeast Asia. Third, as a result of my 1998 fieldwork I had acquired a good understanding of the private production of salt, which is a major source of income in parts of coastal Madura and an industry for long closely linked with Madurese prahu shipping. Fourth, in addition to this background of experience in Madura and prahu circles generally, I had a serious interest in the commodity that has been the economic mainstay of prahu shipping for the past four decades, namely timber. As an amateur carpenter of fair experience — a few years ago I built a ten-metre-long wooden sailing vessel and, as I write, a smaller vessel nears completion in my workshop — the timbers mentioned in this study are not for me abstract commodities. I am familiar with their appearance, their characteristics when worked, their structural properties and, in some cases, even their smell when cut. This familiarity with the common timbers of the region added interest to my field research in 2002–2003 and, as with first-hand knowledge of the traditional vessels of Madura, being able to talk about different timbers with a confidence born of experience was often useful.

I hope that this study will do much to relieve the general ignorance of the world of the maritime Madurese. However, in addition to its relevance for the study of maritime communities in Indonesia, prahu shipping and Madurese society and culture in general, it may also contribute to the growing body of literature on Madurese economic activities and entrepreneurship and take its place alongside such works as de Jonge (1989), on the tobacco industry in Madura, and Nooteboom (2008), on the economic niches of Madurese migrants in East Kalimantan.

Finally, although as a study of a maritime people this book makes frequent reference to particular indigenous vessel types, any reader who lacks background knowledge in this area will not be disadvantaged. The book has been written for the general reader, eschewing nautical jargon that might deter the non-specialist with little interest in technical or other esoteric matters. The story of the maritime Madurese is a significant part of the maritime and social history of Southeast Asia, and I have endeavoured to present it in a way that can be appreciated by a wide readership.

Fieldwork

This study draws primarily on fieldwork carried out in Madura, Java and Kalimantan in 2002–2003, supported by earlier field trips in 1998 and 1999–2000, and drawing on earlier observations and experience in Indonesia going back to 1974. All conversations and interviews in the field were conducted in Indonesian, in which I am sufficiently competent. The names of some key informants have been changed to respect their privacy but place names are unchanged.

The fieldwork was loosely organised and at the start involved mixing with people in maritime communities and ports, observing them and joining them in their work and social activities, and asking about various topics whenever it seemed appropriate. Semi-structured interviews came later, once a broad understanding had been acquired of the prahu economy and the timber trade.

While my field experience provided good opportunity to observe the scope and nature of the contemporary maritime economy of the Madurese, to understand the deeper past of these 'people without history' was more difficult. Although valuable insights in this direction were gleaned from de Bruyn Kops (1854), Hageman (1858, 1862, 1863), van der Kemp (1894), Verwijk (1893), Kuntowijoyo (1980), de Jonge (1989), Knaap (1996), and from some of the reports to the Inquiry into the Diminishing Welfare of the Indigenous People of Java and Madura (1905–1909), written sources alone were not enough to form a coherent picture. Big gaps loomed in my knowledge. Why, for example, were none of the three main types of traditional Madurese prahus of the twentieth century mentioned in de Bruyn Kops' comprehensive 1854 listing and descriptions of prahu types? What had happened to the few large native-owned vessels with European features — schooners and *toop* — listed in the *Regeeringsalmanak* for Sampang, Pamekasan and Sumenep around 1870 (noted in Kuntowijoyo, 1980: 128–9),[1] and why was there no trace at all of these hybrid types, the cream of the contemporary indigenous fleet, among twentieth-century Madurese vessel types? The answers were to be found in the field, but it took a long time before I could discern them.

Critical to unravelling these puzzles was another field resource in the form of the traditional vessels of Madura. In his study of the Parava fishermen of the coast of Coromandel, Patrick Roche reflected, with a harbour scene in his mind's eye, that "there, in the inert log of the abandoned cattamaram, the bounding vallam, and the thonis ringing the cargo vessels, was written the history... of the whole Parava jati [caste]" (Roche, 1984: 193). This sentence is more than just evocative. A similar statement could be made for any of the main maritime groups of Indonesia, but especially so for the Madurese, a people whose traditional vessels were of extraordinarily diverse and distinctive design and who resisted change and outside influences longer than any other of those maritime groups. Nowadays very few Madurese vessels are really traditional, but the changes in this respect have been quite recent, reflecting changes in the perception of the maritime entrepreneurs of Madura about their own material culture and their place in the world. I was fortunate to have had a rich experience of the various traditional vessel types, which, in addition to their economic role, served as cultural markers for the groups who built and used them. As such, they constituted a set of signs informing about local maritime traditions and also, in conjunction with other scattered sources, pointing the way into a past beyond living memory.

For the more recent past, back to the late colonial period, I was able to rely upon living memory through oral history, encouraging older people to talk about earlier periods of their lives and their work in those days. This was not as easy as might be imagined. The economic changes that have taken place in this strand society over the past three decades have been so great that there is now little interest in the past, other than in religious or family matters. What matters

[1] The schooner was a European-type vessel, while the *toop* was similar in build to European vessels, but with an indigenous sailing rig. The *toop* was the prahu type most used by the Dutch for armed patrols (de Bruyn Kops, 1854: 33–7; Paris, 1842: 79 and plate 88).

is the present, especially because people are generally more prosperous now than in the past. Why dwell on times when life was harder? The problems for the researcher in such a situation are illustrated by the following two examples from my own experience.

In 1984 I visited the small tidal estuary of Polagan, near the town of Sampang, and saw a large fleet of traditional prahus on the beach. These vessels were in service, not derelict, and they were based at Polagan, from where they carried salt. But when I revisited this place in 1999, the scene had changed greatly. The tidal flats had almost entirely been reclaimed, leaving only a very narrow channel. There were only a couple of small boats and no cargo prahus. Indeed, there was no space for such bulky craft. I spoke to several local young men about the prahu fleet that had once been based there, but none of them had any knowledge of it. They doubted the veracity of my information until, with historical instincts aroused, I managed to find a man in his fifties who not only clearly remembered the colourful scene that I had witnessed myself 15 years earlier but also provided me with valuable additional information about some of the vessels that used to carry salt from Polagan.

The second example concerns the timber trade. When I was in Madura in early 1998, prahu operators were filled with consternation because of an official crackdown in the East Java port of Pasuruan on the carrying of undocumented timber. This crackdown had begun in November 1997 and such was the dismay in the village where I was based that many people seemed to talk of little else for months afterwards. By 1999, however, prahus calling at Pasuruan were being targeted by officials much less frequently and business was again thriving. Four years later, in 2002, I asked numerous people from the same area about when officials had begun to demand documentation for timber. Remarkably, none of my respondents mentioned the 1997 episode, which seemed to have vanished into the past. Most were emphatic that the tough official stance had begun only in 2002. There was no reason why these persons should have suppressed mention of the 1997 episode, except that it no longer loomed largely in their consciousness because the problem had turned out to be only temporary. Subsequently, I did obtain a better historical overview of the enforcement of the timber transport regulations from informants within the timber industry in Java.

The preoccupation of the people of these strand villages is thus necessarily and overwhelmingly with the present. While this may always have been so, a concomitant of their economic struggle, it has undoubtedly been contributed to by the move away from the traditional vessel types, with all the symbolism and local pride associated with those vessels being abruptly left behind. Nor is there any special place in this society for the raconteur, a type rarely encountered in Madura. The primacy of Islam contributes further to this situation. Throughout Madura there is much preoccupation with religious matters but almost no concern with the profane past, knowledge of which is of no material or social consequence.

Under such circumstances, the insights into the past provided by informants are but glimpses and cannot in themselves constitute a chronological narrative. Nevertheless, by linking these insights with what I learned of Madurese history and historical works relating to prahu shipping and maritime trade, a coherent sense of the past of these maritime communities began to form. In this respect my earlier experience was to prove invaluable, not only in helping me to make sense of statements that might otherwise have remained cryptic but also enabling me to talk confidently about the traditional vessel types and the various ports once frequented by Madurese prahus, thereby encouraging older men to talk about their sailing days.

Even when inquiring into the contemporary timber trade, it was no simple matter to establish an overview. Very few men in these prahu villages have a sound overall knowledge of the timber industry, with its ecological implications and its legal-administrative aspects, or even of where vessels from other parts of Madura obtain or discharge timber. Most are familiar only with their own economic niches. Consequently, the building of a comprehensive understanding of the Madurese involvement in the timber trade was a gradual process that required compiling and verifying the knowledge of many informants from different walks of life and places: vessel owners, skippers, ordinary crew members, boatbuilders, sawmillers, timber wholesalers, local government officials, port officials, shipping agents, forestry officials and water police.

Acknowledgements

Of all the many people who have helped me in this project, none deserves more thanks than Professor Jim Warren, of Murdoch University, Perth. As well as being a mentor of long standing, Jim has been an inspiration to me and it is hard to imagine that I could have made it to the end without his unflagging support. In addition, Professor Howard Dick, Professor Kathryn Robinson, Associate Professor Carol Warren, Dr Gerrit Knaap, Dr Gerben Nooteboom and Emeritus Professor Campbell Macknight have all read various drafts of this study and their constructive criticisms and comments have contributed significantly to the final result. Professor Dick has also been of great assistance in his role as editor of this series, and I suspect that his efforts have gone well beyond what would usually be required.

I am grateful to Murdoch University for the various scholarships I received, without which this study would not have been possible; also for the excellent office accommodation that was made available to me in Murdoch's Asia Research Centre, as well as the moral, technical and intellectual support of my colleagues at the Centre. My thanks are also due to the Indonesian Academy of Sciences (LIPI) and Universitas Muhammadiyah, Malang, for sponsoring my field study in Indonesia. Professor Achmad Habib, my counterpart for the field study, was a stalwart friend at whose house in Malang I frequently stayed. I must also express my appreciation of my wife, Selva, who has had to tolerate my long absences from home and who has supported me enthusiastically throughout this project.

Finally, my gratitude is due to my field informants in Madura, Java and Kalimantan. The people who helped me in the field are far too numerous to all be named here. However, certain individuals stand out and deserve special mention: Haji Amir Hosein and Haji Ariyono, leading vessel operators of Gayam, Sapudi; Haji Rofi'i, the village head of Prancak, Sepulu; Haji Komar, the village head of Sepulu; Sartori, the harbourmaster at Sepulu; Pak Yakob, the former village head of Telaga Biru; Haji Abdul Azis and Pak Sum, vessel operators from Gedugan, Giligenting; Haji Matsahri, a boatbuilder and vessel operator of Batuputih, Sreseh; Haji Rufi'eh, a vessel operator of Taman, Sreseh; Haji Asrani, Haji Zaini and Hamid Sahal, timber merchants in Pasuruan; Hanan, a sawmiller in South Kalimantan; and Abdul Majid, a shipping agent in Juwana.

As an outsider it was often difficult for me in Madura, but the inconveniences and discomforts I experienced were more than compensated for by wonderfully rewarding times when I felt acutely alive and fortunate. My time among the maritime Madurese has been a profound and defining part of my life, a fascination which became an end in itself rather than just an academic investigation. Although this book cannot convey all that I would wish, I hope that it will stand as an adequate record and a fitting tribute to these hardy seafaring people.

1

A Seafaring and Trading People

The homeland of the Madurese people is the large island of Madura, situated a short distance to the north of the eastern end of Java. It is densely populated with a population of over three million, but at least as many Madurese people live on the adjacent mainland of East Java. There are also smaller populations of Madurese elsewhere in Indonesia. Although one of the main ethnic groups of the archipelago, the Madurese are arguably the least understood and, in recent times, the most controversial. Popular perceptions of them amount to little more than a series of negative stereotypes, many of which go back to the colonial period (de Jonge, 1995). These stereotypes have hardly been dispelled by the recent history of ethnic tensions and violence between migrant Madurese and other ethnic groups in West and Central Kalimantan, with bloody pogroms waged against them in these provinces in 1997, 1999 and 2001. The Madurese are widely perceived as an unrefined people with a fierce code of honour, highly mobile across the archipelago, and having an intense determination for material success by whatever means and with scant regard for those of different ethnicity to their own.

Apart from linguistic studies and a few articles about the cattle and fishing industries, academic investigation of Madurese society and culture was for a long time very limited because the island was regarded as just an appendage of Java. The anomaly of this situation was finally recognised in the late 1970s, when a group of concerned Dutch and Indonesian scholars launched the Madura Research Project. Over the next few years many studies were produced under the auspices of this project, some of them outstanding. But despite this surge of interest in the Madurese, nothing was produced about one of the most distinctive aspects of these people, and one setting them apart from other

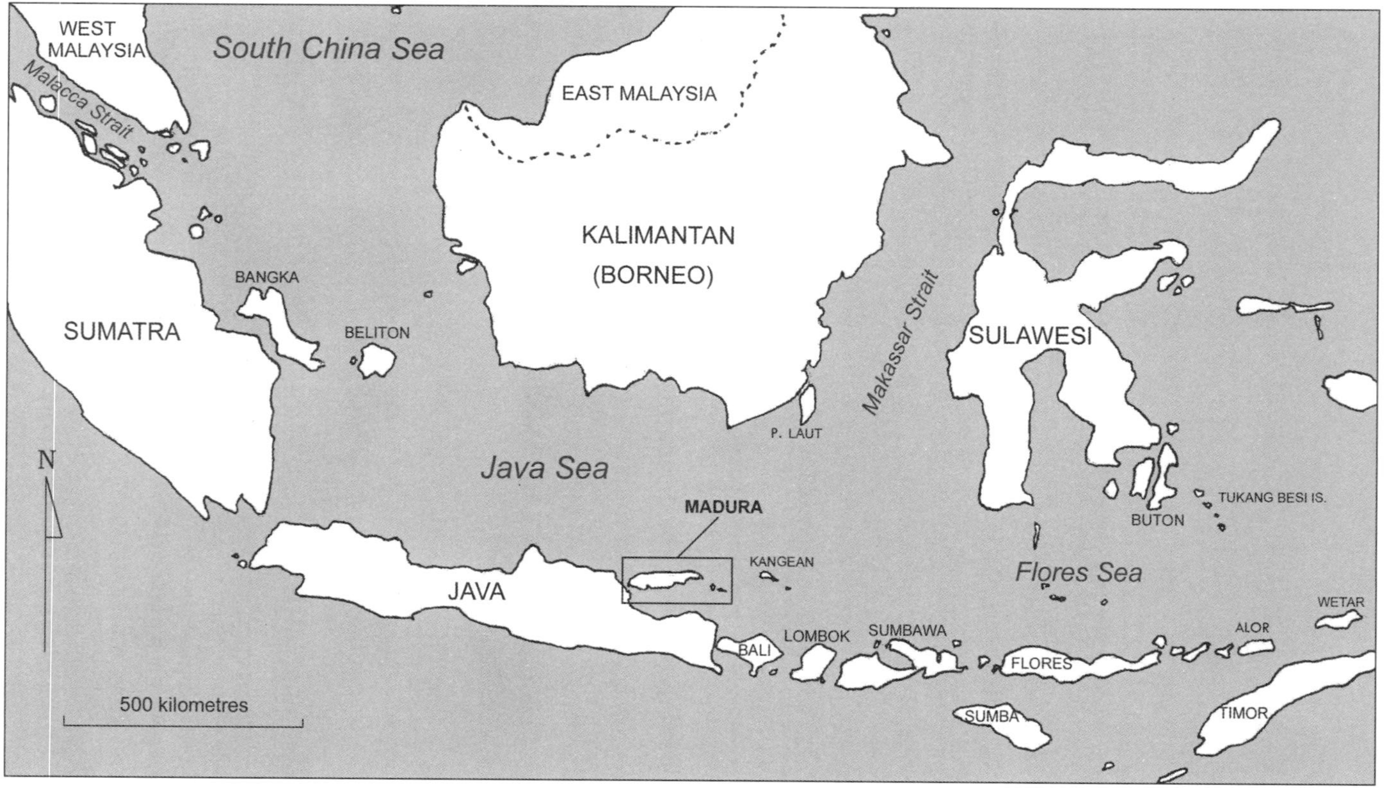

Map 1 Madura in the Indonesian archipelago

ethnic groups of central Indonesia: their role in maritime transport and trade.

A Neglected Maritime History

The Indonesian archipelago has a rich maritime heritage and in premodern times was one of the world's great maritime culture zones (Brissenden, 1976), a consequence of a strong trade demand coupled with very favourable natural conditions. Its waters are extensive, comparable in area to the Mediterranean, and the winds are often fresh with rough seas. However, unlike many other tropical areas, including the neighbouring Philippines, Indonesia is almost free from typhoons, while the monsoonal winds are reliable and conducive to maritime commerce. The main sailing season is the southeast monsoon or dry season, from around late April to early November, but the northwest monsoon was also formerly much used by sailing vessels for eastward passages, especially in its early and late stages (November–December, and March).

Although the pattern of maritime trade that existed until the sixteenth century was severely disrupted by European intrusion and mercantile expansion, vestiges have survived. The most conspicuous example has been the existence of a large fleet of traditional-looking wooden vessels, or prahus,[1] operating in the interstices of the maritime trade conducted by modern shipping. Until about 1980 the majority of these prahus were without engines, constituting the largest fleet of open-water, working, sailing vessels in the modern world. Since the 1980s most prahus have been motorised and their sailing rigs reduced, but a small number continue to operate under sail alone. Although to the casual observer these vessels may all look 'traditional', very few of them still warrant this appellation because of the extensive changes in design that have accompanied the process of motorisation. Yet, regardless of their degree of departure from tradition, none of these vessels are built from plans and they all contrast markedly with modern ship types.

---

[1] The Indonesian word *perahu* is a general term meaning 'boat' and can refer to any sort of small vessel. However, in Madura small open vessels of up to ten metres in length are commonly referred to as *sampan* — although the boats bear no resemblance to the various small Chinese vessels known by this name — while the word *perahu* (Madurese, *parao*) is reserved for larger vessels, usually decked. The word 'prahu' in this work refers to vessels of the latter sort.

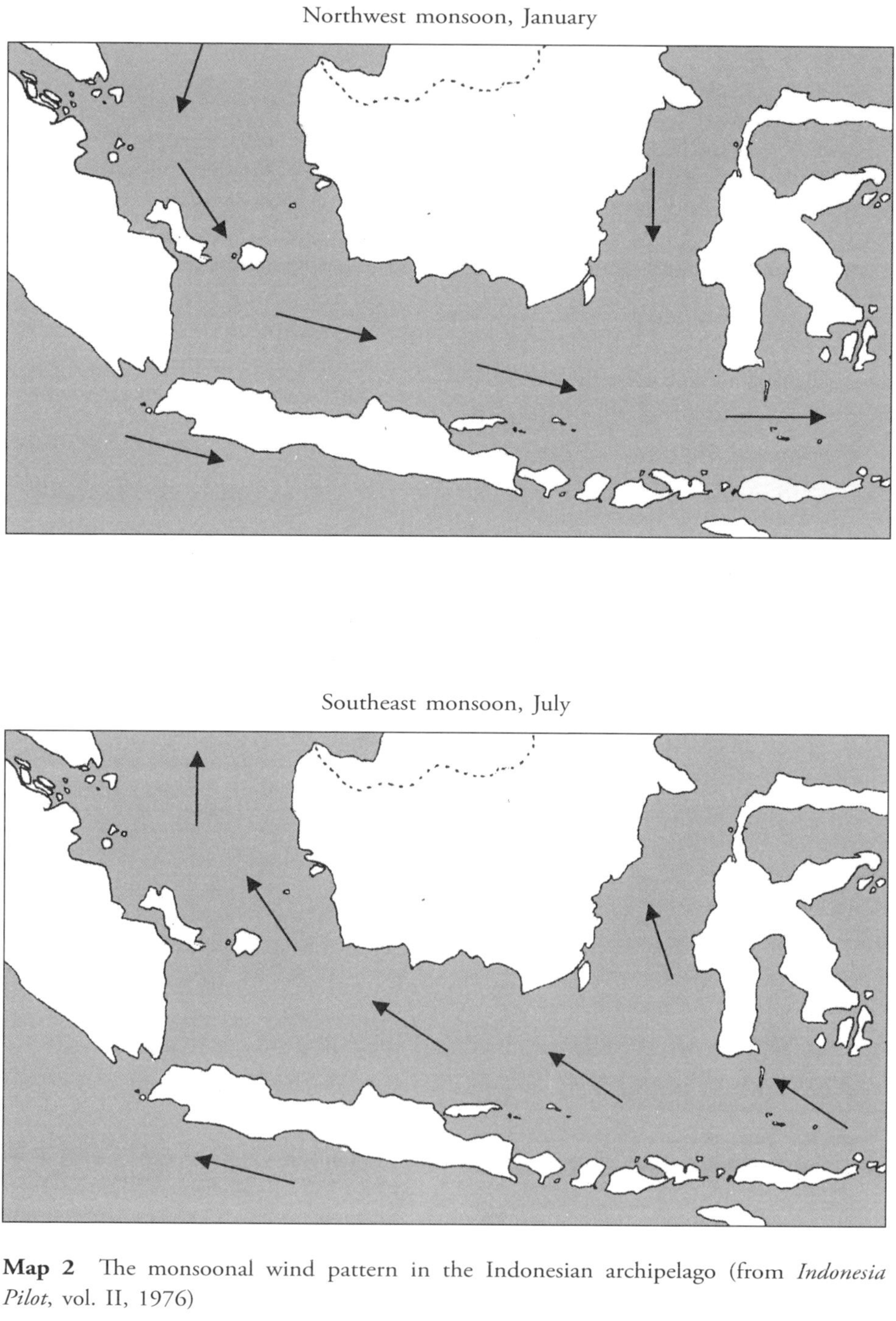

Map 2 The monsoonal wind pattern in the Indonesian archipelago (from *Indonesia Pilot*, vol. II, 1976)

Despite the numerous strand-dwelling communities scattered around the Indonesian archipelago and their different ethnicities, transport prahu activity — the indigenous shipping industry — has been dominated by just five ethnic groups (*suku*): the Bugis, the Makassarese, the Mandar, the Butonese and the Madurese. Of these five maritime groups, the first four mentioned originate from the southern area of Sulawesi, in eastern Indonesia.[2] The people of Sulawesi have long been renowned for their association with the sea, just as the Javanese are associated with prowess in agriculture. What is of special interest here is the apparent incongruity of the inclusion of the Madurese as one of the main maritime *suku* because Madura, being so close to Java, is widely regarded as belonging to the Javanese-dominated core.

By the time of Indonesia's independence in 1945, prahu shipping was already anachronistic by world standards and by the 1970s it was even more so. Nevertheless, the literature concerning modern indigenous shipping in the archipelago is remarkably limited. The two most significant works on the subject are those by Dick (1975a, 1975b), and an unpublished study by Hughes (1984). Both are of a general nature, focusing on the prahu shipping industry as a whole, with Hughes' study following up Dick's long benchmark article to take into account the motorisation of prahus which was proceeding apace after the mid-1970s. Apart from these two primary works, most other studies in this field have focused on the vessels themselves, in most cases with minimal attention to their economic role. Consequently, the representations in the literature of this unique maritime world amount to little more than a series of glimpses, while the motivations, strategies and experiences of the men who actually operate the vessels remain but dimly perceived.

The inadequacy of literature on the maritime tradition and institutions unique to Indonesia is not by chance. Although the major regional centre of power during the first millennium C.E., Sriwijaya, in southern Sumatra, was a maritime-oriented state with its wealth dependent on shipping and trade, the last and greatest political powers that emerged in the archipelago, the states of Majapahit and Mataram, were both agrarian-based with their centres in the interior of Java.

[2] The home area of the Butonese, the island of Buton and smaller nearby islands, is separate from the mainland of Sulawesi but administratively part of the province of Southeast Sulawesi.

Consequently, the Javanese, the dominant ethnic group in the region, perceive themselves as a land-based, agrarian people. Among the status-conscious Javanese, and indeed most of the agrarian peoples of the region, agrarian society is associated with 'high' culture because the principal traditional cultural refinements and institutions sprang from this agrarian base, whereas coastal (*pasisir*) society, with its orientation towards the sea, is regarded as coarse and best avoided.

Thus in Java-dominated Indonesia, there has been an enduring and pervasive bias towards the land and little recognition of the region's maritime past. Indeed, the attitude towards maritime communities has been disparaging. This land bias is reflected in the small number of studies by Indonesian scholars of their own maritime heritage, a field which by virtue of the rich sources available should have been exceptionally rewarding.[3] Foreign scholars have also brought their own bias towards land-oriented societies as a consequence of their own cultural and professional backgrounds. Consequently, even though Emmerson (1980) and Dick (1993) have argued cogently for greater concentration on Outer Island and maritime-oriented societies — with Dick arguing that Madura should for such purposes be grouped with the Outer Islands — Indonesia's maritime communities in general and seafaring peoples in particular have remained neglected by academic researchers. Nowhere has this neglect been greater than in the case of the Madurese.

Of the five main maritime groups, by far the dominant in popular perception as well as the best represented in the literature are the Bugis. In general reference to maritime matters, the term Bugis often implies the Makassarese as well because both ethnic groups originate from the province of South Sulawesi and share a common maritime culture despite their linguistic differences.[4] Although they are less

---

[3] Notable exceptions include Pelly (1975), Lapian (1979), Lopa (1982), Leirissa (1993) and Masyhuri (1996). In recent years interest in maritime history and maritime communities has increased among Indonesian scholars, most significantly at Hasanuddin University, Makassar, and at the Centre for Southeast Asian Maritime Studies at Diponegoro University, Semarang.

[4] Because of the overlap of settlement in the vicinity of Makassar and the acknowledgement of common cultural characteristics, the Bugis and Makassarese tend to regard themselves as a unified people, the 'Bugis-Makasar' (see Pelras, 1996: 14–15). Some sailors on Bugis vessels have in fact identified themselves to me as being of 'Bugis-Makasar' ethnicity. However, because this joint ethnic ascription is not normally applied by outsiders, I have refrained from using it in this work.

often associated with the Bugis, the maritime culture of the Mandar people, from the northwestern part of South Sulawesi, has much in common with that of the other two ethnic groups. Much of what has been written about Bugis maritime culture and history therefore applies also to the other maritime groups of South Sulawesi. An article on prahu shipping of South Sulawesi, by van Vuuren, appeared as early as 1916. Since then numerous other works on this general area have been published, covering historical aspects, vessel development and construction, and navigation.[5]

The maritime Madurese, by contrast, have received only slight attention, with published works in this field limited almost entirely to descriptions of Madurese prahus. The first such work was by Blake (1925), who described one Madurese prahu type in a popular magazine, while Gibson-Hill (1950), Horridge (1981, 1986), and Hawkins (1982) present information and illustrations of traditional Madurese vessels in general works dealing with prahu types from around the archipelago. Mellefont (1991) and Burningham and Mellefont (1994) present more detailed information on other Madurese types and Piollet (1995) provides an excellent pictorial record of various Madurese prahus. But useful as this body of material is, the series of incisive monographs by Horridge (1978, 1979a, 1979b, 1982) on the design, construction and developmental history of the main prahu types across the archipelago has not been complemented by a similarly detailed study of the transport vessels of Madura, notwithstanding the unique designs of the Madurese vessels as well as their economic importance.

Moreover, despite the reputation of the Madurese as seafarers and recognition of them as rivals of the Bugis and Makassarese in seafaring and trading (Dick, 1975a: 74), there has been no significant

[5] Historical studies include those by Tobing (1961), Cense (1972), Lopa (1982), Macknight (1976), Macknight and Mukhlis (1979) and Leirissa (1993); works on vessel types and construction include Nooteboom (1936), Pelly (1975, 1977), Horridge (1979b) and Sularto (1985); while piloting and navigation have been covered by Amarell (1999). Apart from this academic literature, by no means exhaustive, there exist various accounts of experiences of sailing on Bugis vessels, most notably the two books by Collins (1936, 1937), and documentary films, for example *The Hasans: A Buginese Trading Family* (1974). The literature pertaining to the maritime world of the Butonese is less extensive but includes an ethnographic study (Southon, 1995) as well as a study of the Butonese vessel, the European-inspired *lambo* (Horridge, 1979a).

historical or ethnographic study devoted to the development of their maritime capability or their maritime activities, which are frequently alluded to but rarely described in any detail. Dick's long article makes many passing references to the Madurese but almost all of his examples are of the Bugis and Makassarese. Similarly, Lips (1993) does not neglect to mention the Madurese in his study of the development of small wooden motor ships of modern form but traditional construction but his examples are all from South Sulawesi, North Sulawesi, and Riau, while Evers (1983) mentions the Madurese in his article on traditional maritime trading networks of the region but gives no examples of Madurese networks nor any hint of their nature.

The one study to pay close attention to the maritime transport activities of the Madurese is Hughes (1984) which, as noted, is unpublished. Hughes' study was not about the Madurese as such but his fieldwork was wide-ranging. Apart from the busy prahu harbours of Jakarta he also visited three outlying prahu centres: Bira in South Sulawesi, the Tukang Besi Islands near Buton, and the village of Telaga Biru in Madura. The comparison Hughes draws between Bira and Telaga Biru forms one of the highlights of his study. Yet as valuable as his material is on the process of change in the fleet of Telaga Biru, it remains a limited coverage of Madurese transport prahu activity overall and has been read by few.

While serious interest has been shown in Madurese fishing communities and the fishing economy, both by foreign researchers (Jordaan and Niehof, 1980 and 1982; de Jonge, 1989),[6] and Indonesian scholars (Masyhuri, 1996; Koesnadi, 2002), the role of the Madurese in maritime transport and trade has been seemingly invisible for anthropologists and social historians. This neglect is apparent from the comprehensive bibliography of Madura compiled by Smith (2001): of the many titles listed therein, only one, for a brief article by Cordes (1937), contains any reference to maritime transport activities and even there the topic of "freight shipping" stands together with fisheries and cattle-rearing.

With so little published about them and their activities, the Madurese may be regarded as the 'mystery group' of indigenous seafaring in Indonesia. The mystique surrounding them has been

[6] The interest of each of these authors in the local fishing industry was aroused during extended fieldwork in Madura for research projects that were not connected with maritime activities.

enhanced by the reclusive nature of the Madurese sailors, who interact little with outsiders, and also by their traditional prahus, extant until late in the twentieth century. In addition to their striking appearance, the designs of these prahus owed virtually nothing to the outside world, in marked contrast to the obvious European influence in the sailing prahu types operated by the other leading maritime ethnic groups. And whereas the leading Bugis and Makassarese prahu operators have long been based in the port cities of Surabaya, Jakarta, Banjarmasin and Makassar, Madurese prahu owners have continued to be based in their home villages in Madura. Despite the proximity of their home areas to Surabaya, Indonesia's second largest city and the economic hub of the region (Dick, 2002), these prahu communities have remained separate from the mainstream of Indonesian society, and even from the mainstream of Madura. Nevertheless, they have a long history of engagement with the outside world and their village economy has long been affected by, and even dependent upon, external economic developments. The most recent of these external developments, and arguably the most critical, has been the Java Sea timber trade. This study focuses on the role of the Madurese in the timber trade. It presents these seafaring folk not in abstraction, in the isolation of their strand villages, but in the historical context of their involvement with the wider society and economy of Indonesia.

The role of the Madurese in the Java Sea timber trade necessitates that this book also focuses on the commodity itself, timber. Much has been written about the commercial exploitation of Indonesia's forests, especially concerning the political economy of the timber and wood-processing industries, public policy and forest resource management, and social and environmental issues. However, remarkably little scholarly attention has been paid to the movement of timber from Kalimantan to Java for consumption on the domestic market, despite the fact that this trade has for decades been carried out on a large scale.[7] This neglect is not merely one of failure to acknowledge the scale and economic significance of this trade but also, and more importantly, a failure to appreciate the quasi-legal nature of the Java Sea timber trade and how it operates.

---

[7] For example, the excellent annotated bibliography edited by Feretti (1997), covering works dealing with forest resource management, exploitation and community development in Southeast Asia, makes no mention of any study focusing on the Kalimantan-Java timber trade.

Much of what is revealed here about the timber trade, and particularly the interface between 'legal' and 'illegal' timber traffic, applies not only to the Madurese but also to the other maritime groups involved in the transport of timber to Java, notably the Bugis, Makassarese and Mandar. Nevertheless, the Madurese role in the timber trade is of special interest for three reasons. First, by comparison with the other prahu-operating groups, the activities of the Madurese in the timber trade have arguably been more diverse, with a strong involvement in the trading as well as transporting of timber. Second, in the aftermath of the bloody 'ethnic cleansing' of Madurese from Central Kalimantan in 2001 (van Klinken, 2002; de Jonge and Nooteboom, 2006), Madurese vessels were unable to sail to some places where they had previously loaded cargoes of timber because their crews were justifiably fearful of being murdered. Meanwhile, prahu operators of other ethnicity were able to conduct business as usual. For many Madurese maritime entrepreneurs, this was a critical period and some devised remarkable strategies to survive economically. Third, and most significantly, whereas the other groups working the timber trade have discharged their cargoes only in the main prahu ports of Java, some Madurese entrepreneurs started in the 1980s to go outside this system, landing timber in isolated places, mainly in Madura, in order to avoid official scrutiny and the payment of substantial taxes associated with the movement of timber from Kalimantan to Java.

This timber-smuggling strategy of the Madurese touches upon the wider issue of illegality in contemporary Indonesia, a subject which has recently become a focus of academic attention (Aspinall and van Klinken, 2010). The pioneering work in this field is the historical study by Tagliacozzo (2005), which reveals just how extensive was the movement of illicit goods across the Sunda Straits and the boundaries of British Malaya and the Netherlands Indies during the period from 1865 to 1915. The astonishing scale of this smuggling and the large numbers of indigenous actors involved was not because the local peoples were more criminal by nature than peoples in Western Europe. Rather, it reflected the fact that the emerging colonial states and their arbitrary boundaries were largely irrelevant for the indigenous population. Waterborne trade had for centuries been the lifeblood of the Straits area, so from the indigenous perspective, the laws prohibiting the movement of certain goods without large payments to the state amounted to an unreasonable imposition on their traditional right to earn a living by trade. At the same time, these same laws became an incentive for illicit trade because of the greatly increased margin for profit by avoiding taxes.

In the decades after 1915, surveillance by officials became more efficient and the indigenous population of the western part of the archipelago gradually became socialised as subjects of the Netherlands Indies state. In the case of the Java Sea timber trade, however, there has been a strong resonance with the situation described by Tagliacozzo. This is especially so where the Madurese have been involved. Most Madurese prahu skippers are poorly educated and, at least until recently, there was widespread ignorance among them of the timber transport regulations. Until about the late 1980s they were able to conduct their business in their own way on the margins of the state with almost complete freedom from bureaucratic interference, apart from the routine requirements of registration and port administrative business. Since then they have come under much closer official scrutiny because of their failure to comply with the state's timber transport regulations. Consequently, the core of the latter part of this book is in one sense an account of the struggle between these village-based maritime entrepreneurs, struggling to survive in an increasingly competitive and regulated business climate, and officials trying to force them into the fold of the state or, alternatively, to prey upon their vulnerability.

The remainder of this chapter presents material on Madura and its people, and the factors conducive to the maritime orientation of the coastal Madurese.

Madura

The Madurese were undoubtedly a maritime people from ancient times because they must have come to Madura by sea. Indeed the name of the legendary founding ancestor of Madura, Raden Sagoro, means 'Lord of the Sea' (Abdurachman, 1971: 2). Because their language is much closer to Malay than to Javanese, it has been speculated that they came from Sumatra rather than nearby Java (Cribb, 2000: 31). Yet the differences between the Madurese and Javanese extend far beyond language. In order to appreciate the distinctive nature of the Madurese people, it is necessary to know something about the history, ecology and demography of Madura itself. This background will also help the reader to grasp the content of later chapters.

'Madura' here denotes not only the island of Madura but also many smaller neighbouring islands to the south and east, as well as some other small islands well to the north in the Java Sea. The main island is elongated in shape, being approximately 160 kilometres from east to west and averaging about 40 kilometres in width, with its physical orientation parallel to that of Java. It is separated from Java by the Madura Strait, which for most of its length is from 50 to 65 kilometres wide. At its western end, however, the Madura Strait narrows sharply, so that near the city of Surabaya only a few kilometres of water separate Madura from Java. Until very recently, transport across this narrow stretch of water was by ferry. This service was very heavily used, with tens of thousands of people and hundreds of motor vehicles undertaking the crossing each day, so that long delays were frequent. However, in 2009 a high-level bridge, known as the Suramadu (Surabaya–Madura) bridge, was completed across the narrows, greatly reducing transit times.

Apart from its importance for transport between Madura and Java, the narrows is an important waterway and anchorage for general shipping. Until the development of modern port facilities, the ports of Java were mostly shallow and inconvenient, little more than tidal creeks or open bays with no shelter from the northwest monsoon. The great exception was the narrows of the Madura Strait, where the deep and sheltered water provided an ideal anchorage. In the early modern period the main port of Java was Gresik, situated on the narrows. The Portuguese traveller Tomé Pires, who visited Gresik in the early sixteenth century, called it "the jewel of Java in trading ports... where the ships at anchor are safe from the winds, with their bowsprits touching the houses" (Cortesao, 1967 [1944]: 193). Although Gresik became a backwater after the destruction wrought by the army of Mataram in the early seventeenth century, it has remained an important port for smaller ships and prahus, and also has bulk cargo facilities. There is a steady traffic of very small ferry prahus operating between Gresik and Socah, at the southwestern corner of Madura, carrying petty traders and other passengers. As late as the 1980s these very small ferries were traditional Madurese sailing craft, providing a marked contrast with the large, modern, vehicular ferries operating between Kamal and Tanjung Perak. In modern times Gresik has been overshadowed by nearby Surabaya, which has long been the busiest port of the archipelago.

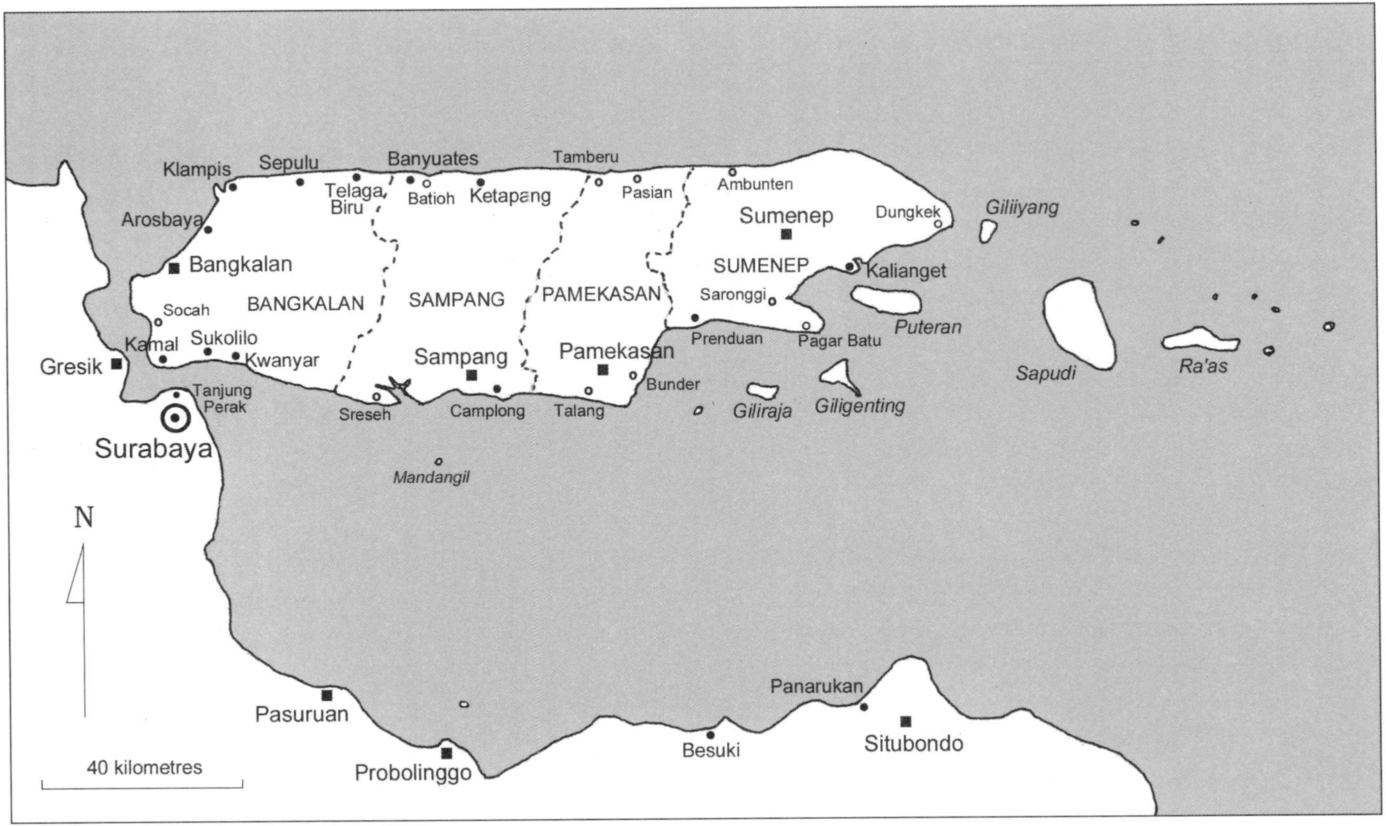

Map 3 Madura, excluding the outer islands of Sumenep

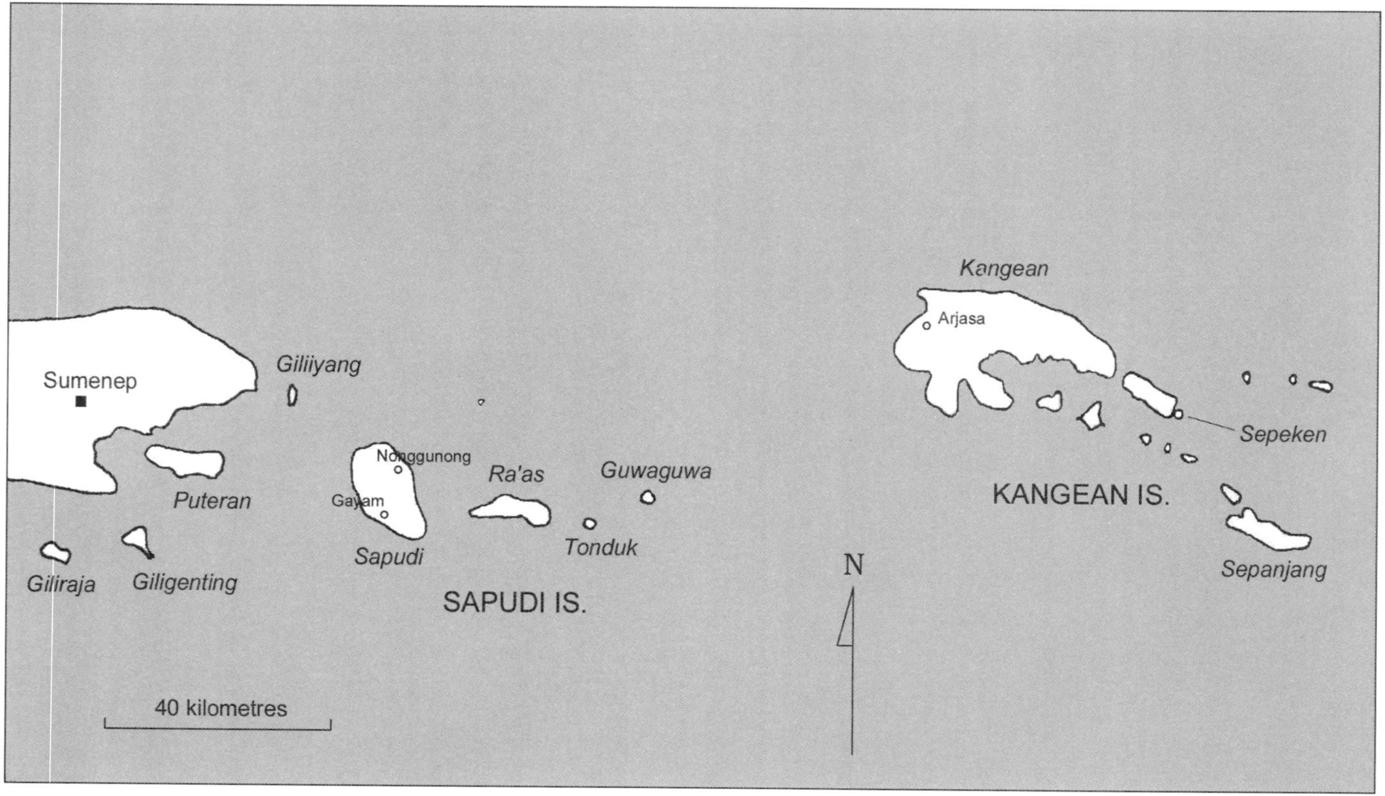

Map 4 The islands of Sumenep, excluding the Masalembu group

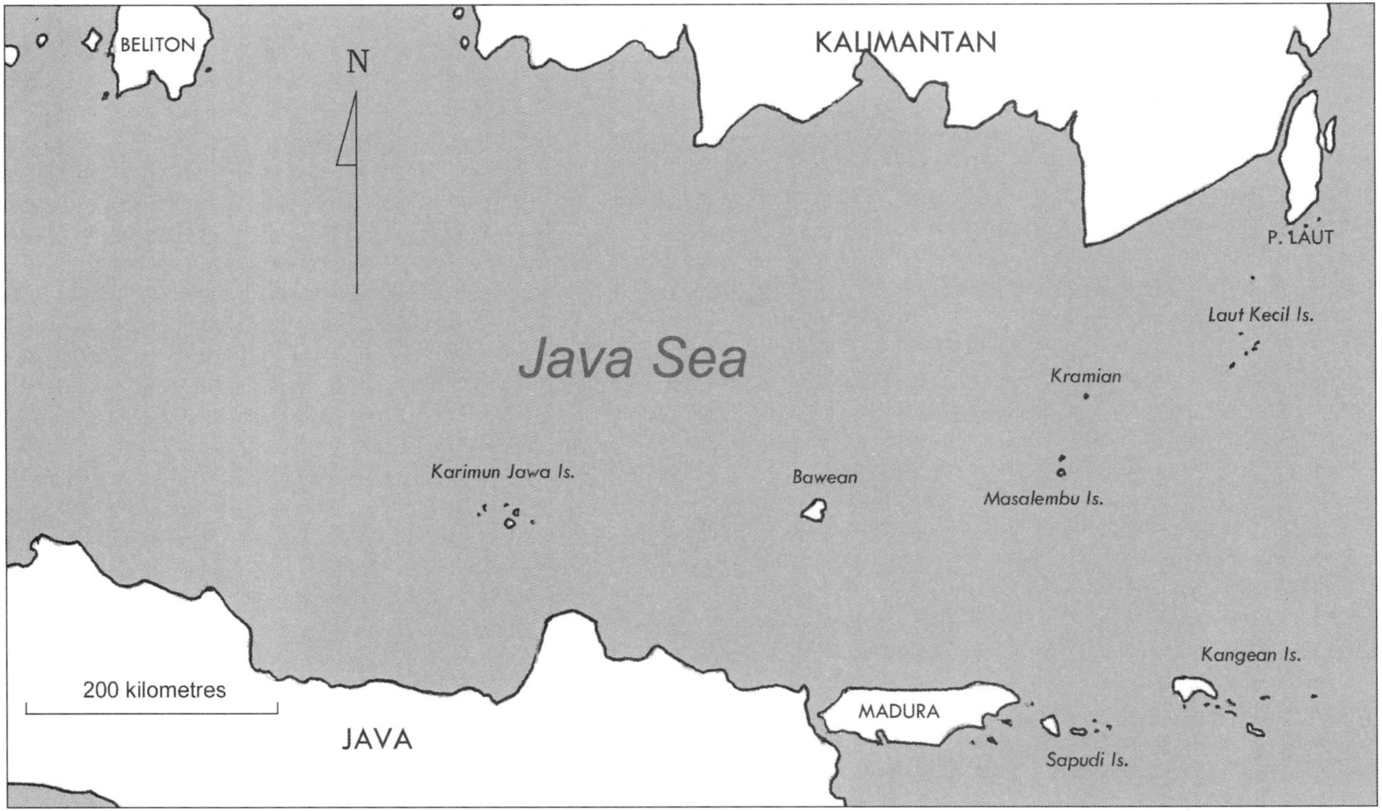

Map 5 Islands of the central Java Sea

Administratively, Madura is part of the province of East Java. The island is divided into four administrative districts (*kabupaten*), being from west to east, Bangkalan, Sampang, Pamekasan and Sumenep. Sumenep is considerably larger than the other three districts because its area takes in numerous islands, including the Kangean group to the east of Madura and the Masalembu group far to the north in the Java Sea. Each district has a capital town of the same name as the district itself and these four centres are the only large towns on Madura. None is any longer a true port town, and none faces directly onto the coast. Nevertheless, they are all close to the coast and situated on the main highway of the island. On the almost dead straight north coast, towns are in general smaller and more widely dispersed than in the southern part of the island. Apart from the ferry port of Kamal, the only significant port on Madura is Kalianget, the port for Sumenep.

In contrast to the spectacular landscapes of Java and Bali, characterised by volcanic peaks, terraced hillsides and lush lowlands, Madura appears unremarkable. It is a low-lying island with its highest point only 471 metres above sea level, although little of it is actually flat. An irregular formation of hills, an extension of those running through the northern part of Central and East Java, runs through the centre of the island and much of the remaining countryside is undulating. Unlike the rich volcanic soils of Java and Bali, the soil of Madura is of low fertility, with much marl (clay mixed with lime) and limestone, and is poorly suited to agriculture. The smaller islands to the east are predominantly limestone. The only significant exceptions to this pattern are the four small alluvial river plains that form the hinterlands and agricultural bases of the four main towns. Considering this geological foundation, it is not surprising that these main towns, each the centre of a former principality, developed where they did. Even so, the fertility of the soil in these alluvial areas is inferior to that of most of Java and Bali.

There are no long rivers in Madura. Only two, the Blega and the Sampang, both flowing to the central south coast, can be regarded as navigable at all, and then only for a few kilometres. Some irrigation is carried out from these rivers, as well as from rivers in the western and eastern parts of the island, but in general irrigation systems on Madura are extremely limited by comparison with Java and Bali. Crops are mainly dependent on rainfall.

In the matter of rainfall also, Madura is disadvantaged by comparison with Java. Although the rainfall in the western part of the island is not much different from that of nearby Java, the central and eastern areas of Madura are markedly drier, especially during the dry season, the southeast monsoon. Combined with the generally poor fertility of the soil and the lack of resources for irrigation, the human significance of this pattern of low rainfall in Madura has been profound. The westernmost part of the island, particularly in the southern approaches to the town of Bangkalan, is moderately well irrigated, so that the rural scene along the main road from Kamal to Bangkalan is not dissimilar from that of much of Java. But this verdant corner is deceptive. To appreciate the typical nature of Madura, one needs to traverse the centre of the island, such as from Ketapang to Sampang, at the height of the dry season when there are only withered stalks in the fields. This was never more than marginal country for agriculture: very uneven with numerous gullies and rocky outcrops, and the naturally thin topsoil prone to erosion in both wet and dry seasons. Yet this same harsh countryside is dotted with small hamlets, their inhabitants eking out a living from such crops as they can grow during the wet season.

Agriculture, Population and Ecology

Irrigated wet-field rice farming (*sawah*) is very much the exception on Madura, so that rice is grown mainly during the wet season. The main form of agriculture is instead dry-field (*tegal*), with by far the dominant crop being maize, which requires less moisture than rice.[8] But whether for maize or rice, the yields per land unit are only about half of those achieved on Java, while the risk of crop failure is higher on Madura (Kuntowijoyo, 1980: 51–2, 97; de Jonge, 1989: 37).

The two main cash crops of the colonial period, coffee and sugar, were both tried and ended in failure. The only agricultural success story from Madura is tobacco, which was introduced in the late nineteenth century and soon became the island's most important cash crop. Tobacco is grown during the dry season, mainly in the eastern half of Madura, with the plants being watered by hand. But although Madurese tobacco is now in demand in the cigarette factories of Java, tobacco farming only became important really on Madura during the

[8] There is some overlap between wet-field and dry-field farming, with rainfall-dependent *sawah* fields often being used in the latter part of the wet season for maize.

1970s (de Jonge, 1989: 155). Most land used to be left fallow during the dry season. Even today, much of Madura's arable land is worked only in the wet season.

Despite the limited suitability of Madura for agriculture, the island has long been densely populated. Indeed, during the nineteenth and early twentieth centuries Madura was the most densely populated large island in the archipelago.[9] Only since the 1940s has Java become more densely populated than Madura. During that decade, marked by the Japanese Occupation and political uncertainty, the population of Madura declined drastically and it was not until 1971 that it recovered to its pre-war level. In 2000 the population was just over three million,[10] representing a 29 per cent increase since 1971.

This large population is predominantly rural, with only about 10 per cent living in urban areas. Moreover, the population is highly dispersed across the countryside. Although large villages exist on the coast and at points along the main roads, the basic unit of settlement is the hamlet (*kampung* or *dusun*), in many cases still consisting of only a dozen or so houses. These hamlets were formerly often occupied by a single extended family, with the dwellings set out in a particular order and surrounded by shrubbery or a bamboo fence to form a compound (*tanean lanjang* in East Madura or *pamengkang* in West Madura). The first head of the compound was the founding husband; upon his death, the husband of the eldest daughter, and so on. New houses were built for married daughters with sons moving out upon marriage to stay with their in-laws. It was a very close-knit social group, with much cooperation and interaction but with each household more or less independent (de Jonge, 1989: 13–17; Muthmainnah, 1998: 22–4; Sudagung, 2001: 53). Although the era of the family

---

[9] At the end of the nineteenth century, Madura was almost twice as densely populated as Java, with 360 persons per square kilometre compared to 219 for Java and Madura combined. For the residencies of East Java at that time, only the 'ricebowl' of Madiun was more densely populated than Madura. The population density of Madura would have been higher still if the large and sparsely inhabited island of Kangean had not been included as part of Sumenep (Kuntowijoyo, 1980: 95; de Jonge, 1989: 21).

[10] Bangkalan, 762,090; Sampang, 715,322; Pamekasan, 792,300; Sumenep, 966,189 (BPS Bangkalan, 2001; BPS Sampang, 2000; BPS Pamekasan, 2000; BPS Sumenep, 2000).

compound has now passed, examples of such compounds incorporated within larger settlements are still commonplace.

The great dispersion of the population of Madura is related to the low productivity of the land. In earlier times, population pressure would lead families to move out of settled areas and clear patches of forest land for themselves. This occurred on a large scale during the nineteenth century, placing pressure on the ecology of the island. Madura had once been largely covered with forest but deforestation took place there more rapidly than anywhere else in the archipelago (Kuntowijoyo, 1980: 32). Because of the excessive clearing of forest, flooding has long been a problem during the rainy season and much of the naturally thin topsoil has been lost to erosion. Moreover, unlike wet rice agriculture, which can be carried out almost continuously with little depletion of the soil, dry-field farming ideally requires that fields be allowed fallow years. However, fallowing is a luxury that the people of Madura have long been unable to afford. Dried livestock manure is burnt as fertiliser but this is not nearly enough to counteract the constant demand on the soil.

Even before this degradation of the land, the Madurese were unable to produce enough for their subsistence. Meilink-Roelofsz (1962: 11) gives an impression of Madura in the sixteenth and seventeenth centuries as a land of plenty, "rich in rice and other foodstuffs", but this is at odds with all the evidence.[11] From the time of first European contact, Madura was a rice-importing area, with the main suppliers being Juwana (Central Java) and Surabaya (Nagtegaal, 1996: 108). In the 1770s, Sumenep was importing around 1,200 tons of rice annually, mainly from Juwana (Knaap, 1996: 113, 230). By the early nineteenth century the deficit had worsened and the Madurese principalities were routinely defaulting on their obligatory rice deliveries to the Dutch (Kuntowijoyo, 1980: 98). Despite the volume of rice imports, rice was the staple diet for only a small proportion of the

[11] Meilink-Roelofsz was possibly relying here on Pires, who noted in the early sixteenth century that Madura was a land which "produces many foodstuffs" (Cortesao, 1967 [1944]: 227). Pires never visited Madura and it is probable that his opinion was based on what he heard and observed in Gresik and possibly Tuban. Both places are very close to the western end of Madura, which has good water resources and is the most agriculturally productive part of the island.

people — the privileged classes and those living in *sawah* areas. For most of the population, maize (an introduced crop) had become the staple. Yet even for maize, the Madurese were far from self-sufficient: in the 1770s Sumenep alone was importing nearly four million cobs of maize per year from the mainland of East Java (Knaap, 1996: 101).

Since the late 1970s dependence on maize has been reduced as a result of the introduction of high-yielding varieties of rice, so that rice is no longer such a luxury. However, these new strains of rice appear to be poorly suited to the upland country of East Madura (Smith, 1995: 161). Maize, often ground up and mixed with rice, continues to be an important part of the diet for many rural households.

History

In pre-colonial times the name Madura referred only to the western half of the island, a principality with its capital at Arosbaya in the northwest, while the eastern half of the island was divided into the two principalities of Pamekasan and Sumenep.[12] Until the seventeenth century, these three principalities, each with its own royal ruler (*rato*), were much like several other principalities scattered along the north coast of Java, all of them under the suzerainty of Mataram with its centre deep in the heartland of Java. This situation changed in 1675 when a Madurese prince named Trunajaya led a rebellion against Mataram. Trunajaya's forces seemed unstoppable until the Dutch East India Company (the Vereenigde Oost-Indische Compagnie, or VOC), alarmed about the threat to its commercial ventures in Java, agreed to help the beleaguered Sultan of Mataram. With the VOC's assistance, Trunajaya was defeated. In the wake of this affair, the Company took possession of the whole northeast coast of Java, which thereafter came under direct Dutch rule. An exception was made for the three principalities of Madura, each of which was permitted, by means of a contract with the VOC, to run its own affairs in return for compliance with certain conditions. The principal factor underlying this special arrangement was the military capability of the Madurese (Cortesao, 1967 [1944]: 227; Nagtegaal, 1996: 61–3; Ricklefs, 1993: 73–6, 91–3).

[12] The Dutch gave the name Madura to the whole island after the establishment of the Residency of Madura in 1857, distinguishing between West Madura (Madura proper, now with its capital in Bangkalan) and East Madura (Pamekasan and Sumenep).

The Dutch made extensive use of Madurese troops to help quell revolts over the next century or so and, in return for this assistance, allowed virtually complete autonomy for the Madurese states after 1830.

Unlike on Java itself, there was thus no disruption of the prevailing social order on Madura. Each of the three states on the island was a classic two-part peasant society, with a large aristocracy supported by a system of tribute from the common folk. The latter did not own any land; at best they had the right to work it. The most productive areas were reserved for the rato, the ruler, while virtually all remaining lands were incorporated into a state appanage system, known as *percaton*, which provided the incomes for the nobility and officials together with their many assistants and retainers (Kuntowijoyo, 1980: App. 6; Touwen-Bouwsma, 1989: 141).

The burden on the common people became worse under this period of self-rule, when the aristocracies became bloated and imposed ever greater demands for tribute and privilege (Kuntowijoyo, 1980: 165; de Jonge, 1989: 76). By the middle of the nineteenth century the abuses were so great that the Netherlands Indies government was obliged to intervene, with the result that in 1857 a Residency of Madura was created with its administrative centre in Pamekasan. In the following year the principality of Pamekasan was abolished, followed by Sumenep in 1883 and Bangkalan in 1885. After the introduction of direct rule for the whole island, it was divided into four regencies with four parallel administrative divisions (*afdeeling*). The new arrangement was based on the old principality boundaries, with the exception of the additional regency of Sampang, which was excised from the territory of Bangkalan.

Although the common people were undoubtedly better off after the abolition of self-rule and the percaton system, their economic situation continued to deteriorate as the population increased. With land utilisation already at its limit, the shortage of food became acute, so that by the early twentieth century Madura was in a state of 'chronic famine' (Kuntowijoyo, 1980: 101). The government belatedly began to plan measures to improve living standards on the island but little was achieved before the Japanese Occupation in 1942.

During the second half of the twentieth century, living conditions on Madura improved, with tobacco especially making a significant difference to the rural economy. It nevertheless remains one of the least developed areas in central Indonesia. By comparison with Java, the number of villages officially classified as backward (*tertinggal*) is high, especially in the western half of the island,[13] while education levels are low with high rates of illiteracy (Sudagung, 2001: 54). It is difficult to foresee much improvement in agriculture given the marginal ecological conditions and the already intensive use of the land. Fishing remains important on the coast but there has been much less development in the fishing industry of Madura than in the centres of Rembang and Pekalongan in Central Java. Moreover, incomes for inshore fishers have declined significantly in recent years as fish stocks in Madura Strait have become depleted (Koesnadi, 2002). Along with fishing, salt production has also been an important enterprise in parts of the south coast of the island, but the future of the private sector of the salt industry is uncertain due to competition from salt production centres in Java and cheap imported salt. As will be shown, the future of the maritime transport sector is also far from assured.

Even the towns have seen little of the development that took place in most of Java after the 1960s. The railway service set up by the Dutch in the early twentieth century ceased to operate after Independence, while ferry passengers arriving at Kamal often experienced long waits for transport to other parts of Madura. However, the recent completion of the Suramadu bridge should change this situation. A property boom was already taking place in Bangkalan before construction of the bridge began and the removal of the bottleneck of the ferry to Java will encourage further investment and development. For a time there was much concern that the ordinary folk of Madura, lacking the skills demanded in an industrialised economy, might be disadvantaged by this development (Rachbini, 1995; Muthmainah, 1998), although it is doubtful that these concerns would have been shared by the many who had to endure frequent crossings on the slow and overcrowded ferry service.

Because of the limited employment opportunities available in Madura and the heavy demands on the ecology of the island, there has long been a widespread dependence on remittances from work elsewhere. In many cases such outside work is seasonal, while in others

[13] During the Sixth Five-Year Development Plan (1994–98) the percentages of such backward villages for Madura were: Bangkalan 68 per cent, Sampang 66 per cent, Pamekasan 30 per cent, Sumenep 33 per cent (Muthmainnah, 1998: 21). In recent years Sampang has been the poorest administrative district in all East Java, with 46 per cent of the population classed as 'poor' (jatim.bps.go.id/, 11/02/2004).

it is indefinite but often retaining close ties with the home village, with the hope of returning there to live after sufficient capital has been accumulated. Often the remittances from outside work are meagre and spasmodic but many families have also managed to save substantial amounts, especially from work overseas. Work in Saudi Arabia is the most lucrative but requires sponsorship from an employer. Such work often takes place in private households — men working as drivers and women, often unmarried, as home helpers. Most Madurese working in Saudi Arabia have a sufficient background in Arabic from their religious instruction to enable them to cope with the language. Many Madurese men have also done well from work in the construction and plantation industries in Malaysia, in many cases illegally. However, this route to prosperity was much restricted after the Malaysian government expelled illegal workers in early 2003.

As a result of work overseas, many families have in recent years built well-appointed houses in rural villages, contrasting sharply with the unpretentious style of older dwellings. As in rural Java, such fine new houses can give a superficial impression of prosperity but do not reflect the economic difficulties facing the majority of households that have not had the advantage of foreign work.

Outward Migration

Disadvantaged by both ecology and historical circumstance, the people of Madura have for centuries been moving from their homeland in search of better opportunities. This outward migration of the Madurese has been the most significant ethnic outflow in the modern history of the archipelago, the numbers far exceeding those of the better-known outward movements of the Bugis and the Minangkabau (Husson, 1997: 98). Not surprisingly, the main destination for migrating Madurese has been the opposite shore of the Madura Strait, the mainland of East Java. During the early decades of the nineteenth century, Madurese poured into the coastal towns of the eastern end of Java in search of a better life. Many found employment in the sugar industry and large numbers later travelled far inland to work in coffee and tobacco plantations. So many came that from 1800 to 1830 the population of the Residency of Besuki increased from an estimated 100,000 to 400,000, with the original Javanese inhabitants becoming outnumbered by 7 to 1 (Elson, 1984: 4). Many parts of East Java, and especially along the north coast, became almost entirely Madurese.

By the middle of the nineteenth century there were as many Madurese in East Java as on Madura itself and the ratio has remained virtually the same ever since (Hageman, 1858: 324–5; *Ensiklopedi Indonesia*, 1983: 2079).

In the city of Surabaya, the impact of Madurese migrants was less pronounced but it was still significant. In 1930 about 13 per cent of the population of Surabaya was Madurese, constituting by far the largest ethnic group after the Javanese, while in the northern part of the city, closest to the coast, the Madurese were in the majority (Dick, 2002: 124, 129). The Madurese presence in Surabaya is conspicuous because many of them obtain their livelihoods from the visible informal sector: as trishaw (*becak*) operators, selling food from roadside or mobile foodstalls (*warung*), or selling a wide variety of goods either as pavement traders or in markets. Many are also employed as guards or nightwatchmen. Madurese communities also exist in the main coastal cities of Central and West Java, as well as many other places around the Java Sea, from Riau in the west to Sumbawa in the east — indeed, almost wherever Madurese vessels sailed to. However, the largest concentrations of Madurese outside East Java are (or rather were, until recently) in West and Central Kalimantan. Many Madurese also reside, temporarily or permanently, in South and East Kalimantan. Madurese migration to Kalimantan has taken place mainly since the 1970s. In most cases this migration was spontaneous, outside the transmigration program of the former New Order government.

Formal and Informal Leadership in Rural Society

Madura is well known as a stronghold of Islam, which is integral with the sense of being Madurese. But Madurese Islam is distinctive because of the special place in society of the local religious leaders, the *ulama* or, as they are popularly referred to, *kiai* (*kiyayi*). Although as a group the kiai form a distinct religious aristocracy with genealogical lines going back to famous early preachers and saints, their special status depends mainly on the power (*ilmu*) that an individual is believed to have. This quality enables a kiai to harness mysterious forces to assist in making predictions, knowing about esoteric matters and, most importantly, influencing the well-being and prosperity (*selamet*) of those around him (Mansurnoor, 1990). Every sizeable village has its own kiai but those with the greatest followings are found in Islamic residential schools (*pesantren*), which in many cases are situated deep in the countryside and well removed from ordinary village life. These charismatic figures constitute an elite and their influence on Madurese society is profound.

The special place of the kiai in Madura is connected with the syncretic form of Islam developed by earlier religious leaders on the island, apparently as a strategy to counter the expanding power of the state and the profane Europeans after the middle of the eighteenth century. Because it accommodated pre-existing indigenous beliefs concerning *selamet* and the spiritual power of ancestors, this new strain of Islam met with ready acceptance across the countryside (Touwen-Bouwsma, 1988). So successful was this approach that when the principalities were abolished in the nineteenth century, the reverence formerly shown to the rato passed naturally to the kiai and continues undiminished today (Mansurnoor, 1995).

However, while the institution of the kiai has enriched the social fabric of Madura and provided the Madurese with an assuredness they might otherwise have lacked, it has not been conducive to economic development. According to Touwen-Bouwsma (1988: 270), the kiai as a group have been the primary obstacle to the economic and political integration of Madura into the Indonesian state. The most obvious example of this isolationist stance was their opposition to the Suramadu bridge (Muthmainnah, 1998). This long-planned project was finally approved only after a senior kiai from Sampang travelled to Jakarta for talks with central government officials and gave his consent to it.

In contrast to the deep deference shown towards religious leaders, the local representative of the state in rural areas, the village-level administrative head (*kepala desa*, also called in Madura the *klebun* or *kliwon*), commands little authority. On Java, historically the village head has been a strong figure, ever since the Dutch appointed tough village leaders ready to use force if necessary to ensure a compliant rural population (Nordholt, 2000). This approach failed in Madura, not only because of the high status of the kiai and their opposition to the colonial state but also, and more importantly, because of the dispersal of the population. The *desa*, the lowest tier of the administrative system, typically includes several scattered hamlets. Every hamlet has its own informal resident leader, who is usually a descendant of the founding family of that hamlet. These informal local leaders can resent an outsider having authority over them, notwithstanding that leader having been elected by the population of the desa. In many cases the klebun commands real respect only in his own hamlet, where he is the accepted leader.[14] This situation has compounded the problems of economic development in rural Madura, especially West Madura (Touwen-Bouwsma, 1989: 150–5; Mansurnoor, 1990: 134–5). Nevertheless, in a few isolated cases, just the opposite applies, with a strong and rugged individual (*oreng blater* or *jago*) being elected to the position and gaining overwhelming support from the local population. These strong local leaders can be a significant force for change in Madura and one such figure features prominently in this study.

Gender Roles and Maritime Culture

Although religious leaders in Madura are always male, women occupy a very important place in Madurese society, well beyond their domestic role as wives and mothers. In two coastal localities that I visited, the desa head was, or had recently been, a woman. In most rural families women play an important economic role by working in the fields, rearing goats or fowls, tending cattle, or selling agricultural produce or fish in local markets. In small villages where there is no marketplace, some enterprising women sell cigarettes and other small items from their homes. In coastal areas, many gather shellfish or prawns, mostly for home consumption, while in one prahu village on Madura women make batik for sale. In recent years many young women, married or single, have earned large sums of money for their families by working as domestic helpers in Saudi Arabia, Malaysia or elsewhere.

Yet, despite their preparedness to do hard physical work, Madurese women are not involved at all in seafaring. Nor are they involved at all in vessel construction, or even the sewing of sails. A prahu is a male domain. Until the 1970s women occasionally took passage on prahus but in more recent times this has not been the case except on prahu ferry services. Wives of skippers do not accompany their husbands at sea. A few widows do own prahus inherited from their late husbands but they take no part in the operation of the vessels.

---

[14] When I was staying in a rural kampung in Sampang in 1998, I was taken on my first day by my host to meet the local kiai. However, my requests to be taken to the local administrative head, the kepala desa, met with no assistance. It soon became clear that in this kampung the kepala desa commanded little respect. It was several weeks before I discovered where he lived, in a hamlet well removed from where I was staying. After I apologised to him for my tardiness in presenting myself, he was friendly and helpful and added that he understood my situation because my host was his "enemy".

External Perceptions of the Madurese

In any consideration of Madurese culture and external perceptions of the Madurese, two points should be noted. The first concerns the need to distinguish between the 'great' tradition of the courts and the aristocracy and the 'little' tradition of the common people. The court culture on Madura was much the same as on Java, and indeed there were close ties between the royal houses on the island and those of Java, dating back at least as far as the thirteenth century (Abdurachman, 1971: 5). Similarly, well-educated Madurese are culturally much the same as their Javanese counterparts and they integrate easily into Javanese society. 'Madurese culture' as a distinct set of characteristics thus refers to the culture of the common people, who are mainly poorly educated and from rural backgrounds. The second point to note is that much of the bad press about the Madurese concerns Madurese from West Madura rather than from East Madura. Most Madurese living in Central and West Kalimantan had migrated from rural areas of Bangkalan and Sampang, where the culture is rougher than in the eastern half of Madura and where education standards are generally low.[15] This is especially so by comparison with the town of Sumenep at the eastern end of the island, which has a more relaxed flavour than the other main centres. More diverse cultural influences are discernible there but in addition the population is generally more prosperous and better-educated than in the western half of the island.

Expatriate Madurese tend not to assimilate into the wider communities around them, although there are many exceptions, especially among the well-educated. Language is a significant factor in this ethnic boundary for many Madurese feel comfortable only when using their own language.[16] Most Madurese, especially younger ones, can speak Indonesian, but often with a heavy accent, reinforcing the stereotype among the Javanese of the Madurese as uncouth by comparison with the Javanese ideal of the refined individual. They are also distrusted

---

[15] As late as the 1980s the majority of Madurese in West Kalimantan had no education beyond elementary school. A study of 400 Madurese respondents in that province found that most were illiterate (Petebang and Sutrisno, 2000: 170; Sudagung, 2001: 75–76).

[16] Davies (1999: 1) notes three 'mutually intelligible' dialects of Madurese: western, central and eastern. However, the differences are only minor and no barrier to communication regardless of where in Madura speakers and listeners come from.

and even feared by many Javanese. In Madurese society the honour of a man, no matter how humble his standing, is a matter of paramount importance. By comparison with standards in Javanese society, offence can be taken very easily and a man may feel culturally obliged to avenge his honour by killing the person causing the offence. This custom of avenging one's honour, usually using the classic weapon of the Madurese, the sickle (*clurit*), is referred to as *carok* (Abdurachman, 1979; Husson, 1991; Latief Wiyata, 2002). Such carok killings were formerly common in Madura, and especially in West Madura, but the incidence has declined in recent decades.[17] The subject of carok has unfortunately often been sensationalised in the press in Java, adding to the mystique of the Madurese and the fear of the Javanese towards them.

Negative images of the Madurese have been reinforced by the recent ethnic violence in West and Central Kalimantan. For many outsiders, the involvement of the Madurese in three major ethnic conflicts — in West Kalimantan in 1997 and 1999 and in Central Kalimantan in 2001 — has confirmed their reputation as an abrasive ethnic group. In West Kalimantan, there had been a long series of lesser violent incidents between Madurese and Dayaks prior to 1997 (Petebang and Sutrisno, 2000: 201) and both the 1997 and 1999 conflicts were sparked by Madurese actions. Nevertheless, in all three conflicts the Madurese were really casualties of political circumstance. In West Kalimantan they were first targeted when long-simmering Dayak frustration over the policies of the Suharto regime finally boiled over. Two years later they were targeted by militant Malays seeking to preserve their domination of the bureaucracy. The violence perpetrated by the Malays on their Muslim fellows served to dissociate them from the disliked Madurese, with Dayak activists soon joining in the killings (de Jonge and Nooteboom, 2006). In Central Kalimantan relations between Madurese and Dayaks seem to have been better, with no long history of violence between the two groups. Hatred of the Madurese in that province seems to have been stirred up by an elite-led Dayak organisation intent on achieving political domination of the province (van Klinken, 2002). In the wake of the 'cleansing'

[17] Large banners bearing the words *Pembunuhan adalah dosa besar* (Murder is a grave sin) used to be hung across the street in marketplaces in West Madura, to discourage the custom of carok. One man whom I had photographed working in a boatyard in Sampang in 1989 was later killed with a sickle in a typical carok case.

of the Madurese from Central Kalimantan, the pre-existing negative images of the Madurese were reinforced, with educated Dayak activists portraying the Madurese as insensitive to indigenous culture, excessively commercially driven and prone to violence at the slightest offence. By comparison, the Madurese, most of whom were poorly educated, lacked any public voice.

Negative perceptions of the Madurese are also evident in the world of prahu shipping, with the difference that these comparisons are mainly with the Bugis. Bugis vessels are generally larger and better-equipped than Madurese ones and the Bugis sailors more outgoing and confident than the Madurese in dealing with outsiders, including port officials. In part this is because the Bugis are more at ease when speaking the national language. The harbourmaster of one of the busiest prahu harbours in Java, a Javanese with many years of experience with both the Bugis and the Madurese, stated to me unequivocally that he preferred the Bugis, whom he respected not just for their seafaring capability but also for their cooperativeness with his office. According to him, not only were many Madurese vessels unseaworthy with a glaring lack of compliance with safety regulations, but many also routinely had big discrepancies in their manifests. However, such shortcomings apply to a great many Bugis vessels as well, especially the smaller ones. This harbourmaster's comparison of the Madurese with the Bugis is in keeping with the observation that whereas the image of most ethnic groups is defined by their best aspects, the image of the Madurese has been defined by their worst (de Jonge and Nooteboom, 2006: 42). Meanwhile, the positive aspects of Madurese culture, such as their piety, thrift and diligence, continue to be under-emphasised.

Maritime Orientation

It should be apparent from this overview of Madura and its people that the Madurese have endured great economic adversity. This prolonged economic struggle has contributed not only to their diligence, thrift and mobility but also to their maritime orientation. Like the Bugis (Pelras, 1996: 3), the Madurese are primarily an agricultural people, but the poor returns from agriculture and the pressure of population were conducive to a coastal drift. The local geography was in this respect advantageous, for Madura's coastline is long and for the most part gently shelving and free of hazards.

The returns from fishing and maritime trade were not especially lucrative but they were significantly higher than that obtained from agriculture alone. The excesses of the appanage system tended to exacerbate this difference, so that during the period of self-rule land ownership was actually a burden on many ordinary folk (Kuntowijoyo, 1980: 361).

According to the Dutch observer Hageman, who was in Pamekasan at a time when the population was suffering under the exactions and abuses of the aristocracy, only the coastal villages offered a decent living standard for the common people (Hageman, 1862: 13). This was reflected in levels of tax paid to the state: in 1871, a typical year, fishing vessel operators constituted the highest tax-paying group in the Residency of Madura, while workers engaged in fishing were also better-off than those working on the land (Kuntowijoyo, 1980: 302, 304). It is therefore hardly surprising that people tended to gravitate to the coastal villages, which became larger and more clustered than the agricultural settlements. In the words of a colonial government report, "The number of households in Madura owning prahus is very great; in the strand villages all the men are seafarers" (OMW, 1907: 131).

This reliance upon the sea for a livelihood is reflected in the popular saying '*abantal ombak asapo angen*' ('for a pillow the waves, for a covering the wind'). In the early twentieth century, many houses in coastal areas of Sumenep used to have finials on their rooftops representing boats (Sulaiman, 1980: 9–11). Similarly, in the administrative district of Sampang (West Madura), one can see banners bearing the slogan *Sampang Bahari* (Maritime Sampang), while a symbolic prahu made from concrete but gaily painted and realistic in shape welcomes visitors to the main town.

Fishing has always been the main maritime occupation. In 1903 there were 8,574 fishing vessels registered in Madura (OMW, 1905: 18). The proportion of the total workforce of Madura engaged in fishing remained constant throughout the twentieth century at around 8 per cent (Kuntowijoyo, 1980: 304; de Jonge, 1989: 41; Muthmainnah, 1998: 19). The many fishing vessels on the Java side of Madura Strait have been operated almost exclusively by Madurese who had migrated to that coast. The numbers engaged in transport prahu work have always been less, especially since about 1980 when engines began to be installed. Hitherto Madurese transport prahus were generally smaller and more numerous, with greater opportunity for crewing work.

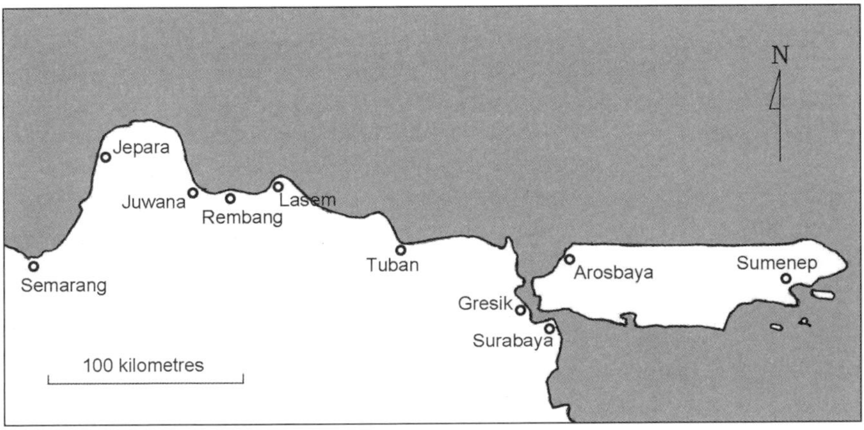

Map 6 Main maritime centres in northeast Java during the early modern period

Although in modern times the Madurese have been the only ethnic group from central Indonesia to play an important role in prahu shipping, this was not always so. When the first Europeans arrived, maritime trade was dominated by Javanese vessels and large fleets were operated from ports along the north coast of Java, from Banten in the west to Gresik (Schrieke, 1966, 19–21; Meilink-Roelofsz, 1962: 270, 272). The most important court and commercial centres of Madura at the time, Arosbaya and Songennep (Sumenep), were two among many centres of maritime activity on this long stretch of coast. The separateness of Madura from Java hardly mattered, for transport was then much easier by sea than by land.

This maritime pattern was disrupted first by the sacking of the north coast centres by the armies of Mataram under Sultan Agung in the late sixteenth and early seventeenth centuries, then by European involvement in sea trade and the restrictions placed by the VOC. Nevertheless, Javanese shipping survived, especially along the northeast coast between Semarang and Gresik. The most important centre for prahu activity was Rembang, which had the great advantage of close proximity to large stands of teak forest. Rembang was also the leading place for vessel construction, with a wide range of vessel types being produced for owners in other parts, including Sulawesi (Schrieke, 1966: 67–8; Boomgaard, 1991: 19).

During the nineteenth century, however, prahu shipping steadily lost ground, first to European-rigged sailing ships and then to steamships. After 1890 this competition became overwhelming with the formation of the Koninklijke Paketvaart Maatschappij (Royal Dutch Packet Navigation Company) or KPM, which linked up all the principal ports of the Indies and monopolised the main shipping routes (Mansvelt, 1938; à Campo, 1993; Dick and Rimmer, 2003: 81–100). There were also railways running the length of Java, transporting goods along the coast more efficiently than by sea, while the price of teak had also risen steeply so that wooden vessel construction was becoming increasingly uneconomic (Boomgaard, 1991: 29–30). By the end of the nineteenth century both Javanese prahu shipping and the traditional shipbuilding industry of the Rembang coast had all but disappeared, and the modern paradigm of the Javanese as land dwellers suited only to agriculture, in contrast to the intrepid seafarers of South Sulawesi, had become established.

The exception to this pattern was in Madura, where small shipbuilding, seafaring and maritime trading patterns not only survived but flourished (Rouffaer, 1904: 50). This is evident from the numbers of trading vessels: in 1903 there were 2,923 trading vessels in Madura, a figure far higher than for any other residency of Java, while in the former indigenous shipping 'capital' of Rembang there remained only 115 vessels. Involvement as maritime transport entrepreneurs was widespread among the Madurese, the number of vessel owners in the residency being nearly as high as the number of vessels, at 2,712 (OMW, 1907: 238). Indeed, by this time observers of the traditional maritime scene were regarding the Madurese as having a special affinity for seafaring, with comparisons often drawn with the Bugis (Kuntowijoyo, 1980: 76).

That the prahu operators of Madura were able to survive into the twentieth century was certainly not because they had been sheltered against competition from modern shipping and railways. On the contrary, they were affected adversely by these developments, not only in Java but also in Madura. From 1901 there was an efficient railway service in Madura, built and operated by the Madoera Stoomtram Maatschappij (Madura Steamtram Company) or MSM. This railway ran all the way from Kalianget to Kamal, with connecting steam ferry services across the narrows to Surabaya. There were also branch railway services between Bangkalan and Blega to Kamal, while various bus and truck services were also established around the island. In addition, the MSM set up a shipping service from Kalianget to Panarukan in East Java, with smaller vessels serving other places along the south coast (Kuntowijoyo, 1980: 291, 293). The competition from these modern transport services on land and sea was felt in prahu centres all around Madura but especially along the south coast of Bangkalan, with a significant decline in the number of large vessels based at the port of Kwanyar as early as 1903. Even as far away as Ketapang, on the central north coast, goods formerly carried to there from Java by prahu were instead being transported overland from the south coast (OMW, 1906: 14, 15). During the 1920s Madurese prahus were also subjected to intense competition from the KPM shipping line, which attempted to drive the prahus off the main shipping routes by slashing freight rates. As well as waging a prolonged campaign of attrition against prahus on the Surabaya–Banjarmasin route, the KPM even tried to capture the trade of Madurese prahus between Sumenep and Banjarmasin (Dick, 1987: 106).

Given that they faced similar competition and difficulties as the vessel operators of northeast Java, the dominance the Madurese achieved in prahu shipping is all the more remarkable. In seeking to explain this success, it is tempting to look to the peculiar situation of Madura under self-rule, long after most of the rest of Java had come under direct Dutch authority. However, the Madurese economy was generally weak throughout this period. Nor were there any large trade centres on the island. Sumenep was the most important place in Madura for trade but it was small compared to Surabaya and Semarang.

In fact, the continuation of self-rule had left Madura somewhat isolated in the region (de Jonge, 1989: 78). Such isolation hardly seemed conducive to the rise of the Madurese as a force in maritime trade. In particular, the preservation of the traditional pattern of social organisation, dominated by the aristocracy and the principality officials, would have militated against the emergence of a middle class of maritime entrepreneurs (cf. Castles, 1967: 4–8). In the late seventeenth century, long-distance trade ventures from Sumenep to Malacca were still predominantly, if not entirely, controlled by the local aristocracy, as was still the case for other centres along the north coast of Java (Knaap and Nategaal, 1991: 131). A century later, it appears that the aristocracy still had a dominant role in maritime trade from Madura, to judge by the 30 or so trading vessels owned by the ruler of West Madura, Cakraningrat around 1780 (Knaap, 1996: 77).

Yet despite this continuing dominance of the Madurese elite in maritime trade, a significant entrepreneurial middle class did emerge, with the most powerful members of this group being Chinese. Among vessel skippers based in Sumenep in the 1770s, Chinese were strongly represented, and may even have been in the majority (Knaap, 1996: 208). With the wealth they acquired through local and inter-island trade, some of these Chinese leased rights to taxation from cash-strapped nobles and became still more powerful. But beside these Chinese and a few Arabs, there were also cases of wealthy indigenous commoners who became 'revenue farmers'. Indeed, the number of these wealthy commoners would probably have been greater if the nobles had not preferred to lease their appanages to Chinese persons (Kuntowijoyo, 1980: 150, 157).[18] Indeed, the very rigidity of the hierarchical social system on Madura may well have contributed to the development of an entrepreneurial spirit among the common folk because success in this direction was their only possibility of social advancement. It is therefore not surprising that some indigenous Madurese became owners of quite substantial vessels. In the year 1871, for example, there were three schooners and a barque based in Sumenep, with one of the schooners native-owned and the other vessels belonging to Arabs; in Pamekasan there were three schooners, two owned by Madurese and the other by a Chinese; and, most significantly, in Sampang there were five *toop* with an average capacity of over thirty tons, all owned by Madurese.[19]

Nevertheless, these wealthy maritime entrepreneurs with large European-styled or hybrid vessels did not survive the competition from modern shipping and railways. Nor did they have any lasting influence on vessel design, for the vessel types which were to dominate Madurese prahu shipping for the greater part of the twentieth century showed no trace of European influence in either hull design or sailing rig. Rather than looking to a wealthy middle class to explain the ascendancy of Madurese prahu shipping over the Javanese, it therefore seems more useful to consider cultural-ecological factors.

---

[18] On revenue farming, see Butcher and Dick (eds), 1993. The most likely reason for the bias in leasing taxation rights to Chinese is that as 'foreigners' these individuals were outside the prevailing social structure. To lease such rights to an indigenous commoner would probably have been regarded as undermining the notion of inherent superiority upon which the whole appanage and tribute-paying system depended.

[19] *Regeeringsalmanak* (1871: 466), noted in Kuntowijoyo (1980: 128–9). The listing of vessels of over ten *last* capacity continued well beyond this date but after 1873 no information is available on the ethnicity of the owners.

In particular, it seems that the prolonged hardships endured by the Madurese had tempered their entrepreneurial spirit. By comparison with the Javanese, their success as entrepreneurs in the late nineteenth century is reflected in the different levels of income tax paid to the state: in 1870, Madurese entrepreneurs (including maritime traders but excluding fishermen) paid on average 2.82 guilders in tax compared with 2.25 guilders for their Javanese counterparts. By 1905 the gap had widened, with Madurese entrepreneurs paying 2.78 guilders compared to the average for Java and Madura of 1.51 guilders (Kuntowijoyo 1980: 285).

This entrepreneurial spirit would naturally have found scope for expression in seafaring and maritime trade, where wit and daring could make a great difference to success. After visiting coastal centres of Madura in the mid-nineteenth century, Hageman described the coastal Madurese as "a bandit in the guise of a fisherman" and "more of a trader and seaman than a farmer" (Hageman, 1858: 346; 1862: 288). Another Dutch observer, writing at the start of the twentieth century, described the Madurese as a "fisherman-shipper" by nature, adding:

> He is much more than a mere landsman like the Javanese. His character is different: he is bolder, with a greater lust for adventure; wilder as well, by comparison with the submissiveness of the Javanese... Formerly notorious pirates — with seafaring a veneer to allow them to go about their business with a semblance of legality — the Madurese have always been a separate type of islander by comparison with the Javanese... (van Deventer, 1904: 109)

Such was the nature of maritime entrepreneurship in Madura during the late nineteenth and early twentieth centuries, dominated not by a mercantile elite but by many rugged individuals who had little choice but to look to the sea for their survival and who were ready to take whatever chances they could get. Like the Madurese in general, they possessed the qualities necessary for survival on the margins of the mainstream economy. In addition, they had two situational advantages over their Javanese rivals. The first was the island nature of Madura and its relationship to Java, with obvious benefits for vessel operators along the south coast and on the eastern islands. The second was the nature of the rural economy of Madura because, as will be shown in the next chapter, the chronic agricultural deficit of Madura provided inherent opportunities for maritime trade.

Overview of Subsequent Chapters

Chapter 2 (Staple Cargoes) focuses on the three most important commodities carried by Madurese prahus, namely, cattle, salt and timber. The coverage of salt and timber has relevance for the material on local prahu fleets in Chapter 3, while the coverage of timber provides important background for Chapter 4, which is concerned with the Madurese role in the timber trade.

Chapter 3 (Madurese Prahus and Their Home Coasts) opens with an overview of the main prahu types used around Madura during the nineteenth century. Thereafter, the greater part is concerned with the three most important types of traditional prahu used by the Madurese during the twentieth century, together with their native coastal areas and their most important lines of work prior to their involvement in the timber trade.

Chapter 4 (Opportunity and Illegality) describes the role of the Madurese in the importation of timber from the Outer Islands to Java, from the late 1960s onwards. Until about 1990, the timber carried to Java by Madurese prahus was undocumented and royalties to the state were routinely avoided. After the authorities began to clamp down on this illegal traffic, Madurese vessels began to discharge their timber cargoes in smaller ports away from official scrutiny, including on Madura itself. Although this strategy was initially successful, an increasingly tough official policy coupled with the introduction of regional autonomy in 2001 has brought serious challenges.

Chapter 5 (Incomes and Profits) examines the financial aspects of importing timber from Kalimantan to Java. Three voyage routes used by Madurese prahus are analysed, showing costs and procedures involved in importing timber, ruses to minimise payments to the state, profits to vessel owners and financial backers, and earnings for skippers and ordinary crew members.

Chapter 6 (Entrepreneurship and Risk) presents a series of individual career profiles which, in addition to their human interest, highlight the risks faced by these village-based entrepreneurs as they struggle to remain viable in an increasingly regulated and highly capitalised shipping market. In particular, despite large increases in vessel size and especially financial outlay, they still they have no access to formal insurance for either vessels or cargo. Those individuals who have been more innovative in their business approach have fared better than those whose business attitudes have remained mired in the past.

Chapter 7 (Conclusion) reviews the main factors behind the success of the Madurese in prahu shipping. This success has not been evenly spread, with entrepreneurs from Sumenep faring better than their counterparts in West Madura and possible reasons for this difference are suggested. Finally, future prospects, taking into account the recently completed Suramada bridge, are considered.

2

Staple Cargoes

Although some prahus from Ra'as in the Sapudi islands were carrying cargoes of trepang, turtleshell and cowries from distant islands like Alor and Wetar to Singapore at least as far back as 1859 (Piollet, 1995: 29), for the great majority of Madurese prahus the cargoes have been more mundane and the outward destinations much closer to home. Indeed, until the boom in the timber trade from the Outer Islands to Java in the late 1960s, the commodities carried by Madurese prahus were mainly goods either from Madura or destined for Madura. In the case of the Bugis, by contrast, their many large prahus were employed mainly in the Kalimantan-Java or Sumatra-Java trades, rather than in the movement of goods to and from Sulawesi.

This chapter provides insights into some of the most important economic niches occupied by Madurese prahus, with an emphasis on the main commodities carried. It opens with a glance at the prahu trade between Madura and Java leading up until the early twentieth century, when the MSM built its railway system running the length of Madura with connecting steam ferries to Java. Before the development of the MSM's modern transport network, prahu traffic across the Madura Strait had been heavy for it was, above all, upon the demand for the transport of goods to and from Java that the maritime capability of the Madurese was founded. The greater part of the chapter then focuses on three of the most important commodities carried by Madurese prahus: cattle, salt and timber.

Historically, cattle and salt have been the two main export products of Madura and they have both been so important as prahu cargoes that the history of Madurese prahu shipping cannot be understood without reference to them. These two commodities are therefore considered separately and in detail to show how they became so important in the rural economy of Madura and their significance for the development of local prahu shipping. Their importance as cargoes will become more apparent in Chapter 3, where it will be seen how the movement of cattle and salt enabled some local Madurese prahu fleets to survive while prahu fleets elsewhere withered.

The last of these three main commodities considered here, timber, is less directly connected with Madura because, as a prahu cargo, it has been carried mainly from Kalimantan to Java. Nevertheless, as the mainstay of the prahu economy from the late 1960s until the present, this commodity is given the same detailed treatment, showing the factors behind the rise of the Java Sea timber trade and how prahus were able to dominate in the movement of timber for the domestic market despite the intentions of the central government. Madurese prahus were far from the only prahus in this trade but, as will be shown in Chapter 4, they were able to establish their own niches, making them a special case for consideration.

The Maritime Trade Relationship with Java

In the early modern period, most long-distance trading ventures from Madura were financed by members of the aristocracy. Vessels from Sumenep were especially well represented, with all the leading nobles of that principality having skippers, referred to as *panakawan* (servants), permanently in their service during the first half of the eighteenth century. On their outward voyages these prahus from Sumenep carried salt and on their return voyages they would bring rice, which was always in short supply in Sumenep (Nagtegaal, 1996: 113). These voyages to places as far away as Malacca, Palembang, Pontianak or Singapore required capital resources which few others could muster. However, with the decline of the old regime and eventually its abolition, wealthy commoners became increasingly involved in long-distance trade. Nevertheless, in overall terms it was trade with Java, and especially across the Madura Strait, that was the most important.

The underlying force driving this trade across the Madura Strait was the same as that for long-distance trade: the inability of the Madurese to grow enough food crops to feed themselves. However, unlike the case with long-distance ventures, the owners of vessels engaged in this short-haul trade usually lacked the resources to become large traders. Instead, their vessels carried groups of small traders and their wares. These small traders were mainly enterprising peasant women who would buy produce from their neighbours to sell in the markets of Java, often for minute profits (Nagtegaal, 1996: 108). Some would then buy rice or maize to sell on their return.

The main destinations in this short-haul trade were Situbondo, Besuki, Probolinggo, Pasuruan and Gresik. So dense was this traffic in the Madura Strait that in the early twentieth century, for the port of Probolinggo alone there were over 4,000 trading prahus arriving each year (van Deventer, 1904: 208). There was also a good deal of maritime trade between the three principalities of Madura because, until the development of modern transport systems, carriage of goods by sea was much easier than by land.

The goods exported from Madura to Java were notable for their variety. In the 1770s the main export goods from Bangkalan and Sumenep were fish, timber (teak), textiles, mats, oil (unspecified), tamarind, dyewood and hides (Knaap, 1996: 228). By the middle of the nineteenth century, Madurese prahus were carrying a wide range of export goods including textiles, garments, umbrellas, cotton yarn, kapok, dried fish, peanut oil, rope, firewood, bird nests and livestock. On the return voyage from Java, many consumer goods were imported from Java together with the staples of rice and maize: earthenware, tobacco, gambir (for tanning hides), clothing, cotton yarn, dye, soap, tea and coffee (Kuntowijoyo, 1980: 125).

The vessels used in this regional trade were not large. Knaap's observation that maritime trade along the north coast of Java during the 1770s was carried out by a "mosquito fleet" (Knaap, 1996: 37) seems particularly apt for Madura. For example, on one of the most important trade connections with Madura, the Juwana–Sumenep route, the average volume of rice carried per vessel bound for Sumenep was only 132 *pikul* or about seven and a half tonnes, significantly less than the typical cargo in the rice trade along the north coast of Java at that time (Knaap, 1996: 113). More than a century later the size of Madurese prahus seems hardly to have increased — in 1903 the average listed capacity for Madurese prahus was only about 16 cubic metres.[1] Yet although the vessels used were small, with correspondingly modest profits per voyage, their combined economic role was very significant. In the words of the social historian Kuntowijoyo (1980: 76), "sea transportation had long been the lifeblood of the Madurese".

[1] Calculated from data in OMW (1906: 20).

With the development of the railway network on Madura and connecting ferry services to Java, this reliance on prahu shipping for petty trade across the Madura Strait all but disappeared. The only exceptions were at Socah, at the southwest corner of Madura, from where petty traders could take small prahus to Gresik, and at Talang on the south coast of Pamekasan, from where traders and passengers could take a larger prahu to Besuki in Java. In both of these cases, the description by Nagtegaal (1996: 108) of waterborne local trade in the late eighteenth century would still have been apt as late as the 1980s. Even the vessels would not have been out of place two centuries earlier. Elsewhere, the role of prahus in the movement of commodities to and from Madura shrank to areas that modern transport could not easily serve, carrying mainly bulk construction materials such as sand, slaked lime, coral stones, roofing tiles, bricks and cement. However, the two commodities most strongly identified with Madura — cattle and salt — continued to be carried by prahus for much longer.

Cattle

Although animal husbandry, especially of goats, is widespread around the Indonesian archipelago, the cattle-rearing industry of Madura is a special case. Indeed, one colonial stereotype of the Madurese was that the ordinary man on the island "loves his *sapi* [cow] more than his wife" (de Jonge 1995: 17). This of course reflected not the true nature of conjugal relationships in Madurese households but rather the importance of cattle in the rural economy. The cattle-rearing industry of Madura was reputed to be the best in the entire archipelago and beef livestock was the most valuable export earner of the island (ENI, 1917–1927, vol. II: 635). This may seem incongruous because Madura lacks the extensive pasture lands associated with the cattle industry in more affluent countries.[2] Nevertheless, the importance of cattle on Madura is related to the ecology of the island.

[2] The Englishman G.W. Earl, who visited Sumenep by ship in 1832, mentioned Madura's fame as a cattle-producing area, together with its "rich pastures" (Earl, 1971 [1832]: 77). However, Earl's experience of Madura was limited to Kalianget and the town of Sumenep, and his view in this respect reflected his preconceptions of cattle farming rather than the local reality. Moreover, his visit was in early November, a time of year when the land would typically have been parched.

Open country ideal for farming cattle is rarely available in Indonesia, particularly in Java. Where conditions are suitable, wet rice agriculture has generally been the traditional economic activity of choice, with livestock rearing mainly limited to buffaloes necessary for ploughing. However, in areas not suitable for wet rice agriculture, the raising of beef cattle was a more attractive prospect, one which could often be combined to advantage with dry-field crop farming. Thus in the early seventeenth century the main livelihood for the residents of Tuban in northeast Java, a place with poor water resources, was the rearing of cattle and horses (Meilink-Roelofsz, 1962: 284; Schrieke, 1966: 20). The same would undoubtedly have applied at that time for many places in Madura. There was still enough free land then to allow cattle to graze on open land away from the villages during the daytime and in the late afternoons they would be led back to their pens adjoining the owners' houses. But such an easy method has long been unavailable to the Madurese.

In the context of modern Indonesia, it is a characteristic of cattle as a commodity that lack of pasture land can be compensated for by diligence and energy on the part of the owner or carer. In this respect the Madurese excel. Their cattle usually remain penned up alongside the owner's house, with the owners and their assistants, often children, having to fetch the large amounts of water and fodder that these animals need every day. This work is monotonous and sometimes arduous but it is as routine and unquestioned as feeding one's own family and well justified by the economic returns. Cattle were traditionally second only to land as an economic resource on Madura and it has been argued that in the nineteenth century, the buying and selling of cattle at the local markets may have been the main source of finance for the importing of foodstuffs into Madura (Kuntowijoyo, 1980: 354). The importance of cattle in the rural economy has barely changed since then: Glenn Smith, who conducted fieldwork in upland Sumenep in the 1980s, noted that almost every household in the locality where he was based either owned or tended cattle (Smith, 1989: 279).

The enthusiasm on Madura for cattle reflects much more than just the sale price for slaughter. Unlike the case for most of Java, cattle-raising complements the prevailing dry-field ecotype and can be regarded as part of a holistic farming system (Smith, 1995). First, the dried maize plants (that is, all parts other than the cobs) are a valuable source of fodder in the dry season. Second, it is advantageous either to own a pair of cattle or to have the right to use a pair of cattle by tending them for someone else. In East Madura, because of the long dry season and the unreliable rainfall in the wet season, it is vital to plough the fields as soon as the first rains have fallen. Delay in ploughing can make the difference between crop success or failure, or getting in a second or third crop late in the wet season. Buffalo could be used for ploughing but the local cattle are better suited to the dry conditions and heat and require much less food and water. Furthermore, if the draught animal is a cow, as is mostly the case, it requires less food and water than a bull, is easier to control,[3] and has the further advantage of producing offspring. Even if one does not own a field, or have access to one, it is still possible to own or tend cattle. Many fishermen keep cattle penned alongside their houses, and when grass is available the animals can be tethered during the daytime along the shore. It is not necessary to raise cattle to full maturity and it is common for an animal to be sold two or three times to different parts of Madura before being finally sold for slaughter. Young male cattle from Sumenep, where gathering food during the dry season can be difficult,[4] are often sold to owners in Sampang or Bangkalan for fattening. On cattle market days the streets in the cattle market towns can be so blocked up with cattle and vehicles for transporting them that it can be difficult to make one's way through the throng, even on foot.

The Madurese do not themselves eat much beef. It can be bought in large markets, but such meat as is eaten is mainly goat mutton, with beef reserved for special occasions. By far the main market has been Surabaya, followed by the other coastal towns of East Java, especially Pasuruan and Probolinggo. Cattle were formerly carried by prahu to these towns from centres along most of the southern coast of Madura, from Kamal to the eastern islands. Cattle were also exported in smaller

[3] By comparison with most domesticated bovines, Madurese cattle are temperamental creatures. Cows outnumber bulls by at least two to one across the island, while in Sumenep, the driest area and thus the most difficult for obtaining fodder, the ratio is about five to one. Given this preference for cows, Smith suggests that the well-known bull races of Madura may have been devised as a means of ensuring a good supply of high-quality breeder bulls (Smith, 1995: 165).

[4] In the islands to the east of Sumenep the problem of obtaining fodder can be so great that according to Smith (1989: 281), farmers sometimes transport their cattle by prahu to the mainland of East Java to let them graze along the shore. When the animals are replete, they are loaded on board and carried back home.

Photo 1 Ploughing after the first rains on the north coast of Sumenep. While the ploughing was being done, a child was following, planting maize seeds.

numbers by prahu to Sulawesi, Kalimantan, Bangka, Beliton and other places across the Java Sea. This export trade in cattle was controlled by indigenous Madurese merchants and was one of the few important business lines not dominated by Chinese entrepreneurs. The trade was at its peak at the end of World War I, when over 100,000 cattle were being exported annually (Kuntowijoyo, 1980: 358).

The transport of cattle to East Java was dominated by prahus. Nevertheless, they faced tough competition from the rail and steamship service operated by the MSM, with the prahu operators further disadvantaged by bureaucratic discrimination against them. Prahus departing Kamal for Surabaya had to pay clearance fees to obtain a departure permit, while the MSM vessels were exempt from this requirement. The time wasted in this procedure was critical. Apparently it was contrived so that the prahus carrying cattle would arrive late for the Surabaya cattle market, whereas the MSM vessel was always able to arrive in good time. Yet despite this discrimination and the efficiency of the new services, prahu shipping managed to survive along the south coast by undercutting the rates charged by the MSM. Consequently, until 1937 the transport capacity of prahus from the towns of Kertasada (near Kalianget), Prenduan, Pamekasan, Sampang and Bangkalan exceeded that of the MSM (Kuntowijoyo, 1980: 296–7).

Since the economy began to move ahead in the late 1960s, the role of prahus as transporters of cattle has steadily declined. In recent decades virtually all cattle exported from Madura to Java have been carried by truck, using the Kamal vehicular ferry, with those from the eastern islands being transshipped via Kalianget. Although some livestock, cattle and goats are still exported from the north coast to Kalimantan, few vessels are now involved in that trade. As will be shown, however, the economic staple of cattle transport was a precursor to the involvement of many Madurese prahus in the transport of timber from the Outer Islands to Java.

Salt

Although salt is nowadays a cheap and commonplace item, it has been of immense political and economic significance in human history. Its value for human society has been as a preservative and as a condiment or something to be consumed in its own right, with the latter use the more important in the Indonesian region. Despite the warm and humid climate, the use of salt as a preservative was less vital than in Europe and North America for the reason that fresh food was more regularly available throughout the year. As a preservative, its main use was in the fishing industry.[5] The salting of meat in the archipelago was a minor industry because Indonesians ate much less meat than Western European people, although in Madura salted beef was formerly a valuable source of income.[6] There were no butter or cheese

[5] Although salt was used for the preserving of fish in Southeast Asia, less salt was required per unit of fish than in far northern fisheries because of the efficacy of sun-drying in the tropics. With small fish it is enough to slice them open, dip them into seawater and leave them on the beach for two or three days. By comparison, oily northern fish such as herring required a lot of salt to preserve them and they also needed to be packed in barrels (Kurlansky, 2002: 114, 139–40).

[6] G.W. Earl noted in 1832 that "large quantities" of beef were cured on Madura and sold for export, and that this product was prized by indigenous peoples around the archipelago. Earl himself regarded this cured beef of Madura as vastly superior to South American 'jerked' beef. But as with dried fish, this curing process did not require very much salt: the meat was simply dipped in brine, coated with pepper, and left to dry in the sun (Earl, 1971 [1837]: 77).

industries, both of which require salt. Salt was needed in the curing of hides, but this was not a large industry.

Nevertheless, salt was very important in the Indonesian archipelago as an item for human consumption. Many foods taste better with the addition of a little salt. Moreover, the body needs a certain intake of salt, especially in tropical regions where so much salt is lost through perspiration. A moderate consumption of animal protein meets this physiological need, which is presumably why gatherer-hunter societies did not make or trade salt (Kurlansky, 2002: 9, 201). Probably the extraction of salt for human consumption was a consequence of the shift to agriculture, vegetarian diets being rich in potassium but low in natural salt. The diet of Indonesian peoples was very low in animal protein which, combined with the tropical climate, led to a high demand for salt.

In Indonesia, salt could be obtained only from seawater, whereas the largest concentrations of people, being agrarian, lived inland. Responding to this constant demand for salt, strand peoples across the archipelago became involved in the production of salt, not only for their own needs but also as an item of barter with inland villages. Indeed, in pre-modern Southeast Asia, salt was the main imported item for inland villages (Steinberg: 1987: 50). As an indication of the importance of this commodity for inland peoples, in the mid-nineteenth century there were an estimated 300 vessels involved in the shipping of salt from the north coast to the interior of Java via the Solo river, while in Borneo, where salt was particularly scarce, some Dayak people were reported to have paid up to 20 units of rice for a single unit of salt and to have used the salt-biscuits they purchased as a form of money (van der Kemp, 1894: 134, 228).

In many cases the traditional trade in salt was of a local nature, as in the Solor Islands to the east of Flores, where strand people have for centuries produced small amounts of salt for trade with upland communities (Barnes, 1993). In Timor, salt has been a means of barter between coastal and inland people since ancient times (Ormeling, 1956: 25, 116). Elsewhere, salt transcended purely local trade. In the Sulu region of the southern Philippines, some Iranun and Samal communities were involved in salt-making on a large scale in the eighteenth and nineteenth centuries, and the sale of salt in hinterland areas became an important line of business for Tausug traders (Warren, 1981: 70–1, 298). Yet across island Southeast Asia, nowhere was more important for salt production than the northeast coast of Java and Madura.

The reason for the importance of this particular stretch of coast for salt production was that it was so well suited to the extraction of salt by natural evaporation. In most other places across the archipelago, salt was obtained by boiling seawater — an inefficient method that produced a small amount of salt at the expense of a large amount of firewood. While such techniques were acceptable for subsistence and local barter, it was only by natural evaporation that large surpluses could be achieved. Yet high productivity required more than just access to a coastline. It also required a broad expanse of flat land almost at sea level, soil highly impervious to water and prolonged spells of dry weather. These ideal conditions for the winning of salt existed along the northern littoral of Java, about which Stamford Raffles wrote: "nearly the whole of the north-east coast of Java and Madura abounds with places well calculated for its manufacture, and unfit for any other useful purpose..." (Raffles, 1978 [1817], vol. I: 177).

The main traditional salt-producing areas in Java were Banten, Krawang (to the east of Jakarta), Cirebon, Semarang, Rembang, Gresik and Surabaya, while on Madura, Sampang, Pamekasan and Sumenep were important centres of production. In each place the local ruler claimed a monopoly on salt, although in practice this was farmed out to a merchant with the salt producers given one-third of the sale price and the ruler and salt farmer keeping the rest. The salt farmers, most of whom were Chinese, became local overlords, with authority over extensive rice-producing areas as well as the salt lands. They sublet the sale of the salt to other Chinese merchants, who in many cases charged such exorbitant prices as the market could bear (Raffles, 1978 [1817], vol. 178). Because it gave a substantial income with minimal complication, this farming out of salt contracts was continued by the VOC after the Company took control of northeast Java in the late seventeenth century, and later by the early Netherlands Indies state (Knaap and Nagtegaal, 1991: 142). However, during the period of British rule (1811–16), the government was determined to bring the sale and distribution of salt under control so as to protect the welfare of the population and, more importantly, to boost its own revenues. In 1813 it abolished the contract system and decreed a state monopoly on salt, with all salt thereafter to be sold only to the state. This monopoly was retained by the Dutch after their return in 1816, although between 1829 and 1847 there was a partial return to the contract system and much debate over the merits of the state monopoly (van der Kemp, 1894: 1–11; ENI, 1917–1927, vol. IV: 865).

Prior to the imposition of the monopoly, the centres for salt production of principal concern to the Dutch were those of Rembang and Semarang rather than Madura.[7] Only in the 1820s did the government begin actively to be concerned with salt-making on Madura, dealing directly with producers, supervising production and exempting workers from the requirements of corvee or service labour in order to have an adequate work force for the industry throughout the dry season (Kuntowijoyo, 1980: 184). Nevertheless, the salt industry remained widely dispersed along the north coast of Java as well as on Madura.

In the second half of the nineteenth century, however, Madura became increasingly important for the government's plans. The catalyst for this was a serious shortfall in the supply of salt from 1859 to 1865. To cope with this 'salt crisis', the administration ruled in 1861 that the salt lands of Sampang, which had been closed for 30 years in order to prevent a glut, be reopened. With the crisis over, the debate over the advantages of the monopoly by comparison with the leasing of salt rights resumed. In 1868 the directors of the salt industry declared that under no circumstances should control be relinquished over Madura's salt production areas because those areas produced the best-quality salt and needed to be retained as a reserve against all contingencies (van der Kemp, 1894: 266). In 1870 Madura was designated as an area for the concentration of the industry and this decision was given force with the issuing in 1882 of an historic ordinance assuring the salt monopoly. As well as reaffirming the government's monopoly on salt purchases for fixed prices, this new law banned the making of salt except with a licence, these being issued in practice only for Madura.

Salt production was limited to the south coast of Madura, in three main areas: the lower reaches of the Sampang and Blega rivers in Sampang; at Bunder and nearby Tjandi in Pamekasan; and the

[7] From the 1680s to the 1740s, the VOC had tried to establish a monopoly on the sale of salt to the west coast of Sumatra to force the inhabitants of the interior of that region to trade their pepper and gold with the Company. Most of the salt purchased for this purpose came from the Rembang area, and Rembang and Semarang were the first areas for which salt-farming rights were auctioned by the VOC, in 1744. In the case of Madura, the farming of salt remained a prerogative of the local rulers until the abolition of self-rule, although the Dutch did lease the salt lands in the latter part of the self-rule period (Knaap and Nagtegaal, 1991: 133–47).

land around Sumenep Bay. Some salt was made on the western part of the north coast but that area is less suited for salt production than the south coast centres, which have three distinct advantages. First, they each have extensive areas of flat land almost at sea level. Second, the eastern half of this coast is very windy during the dry season. The southeast wind, called in Madura the *gending*, loses most of the little moisture it carries in the mountains of East Java and blows strongly across the Madura Strait. Nowhere along the north coast of Java is so windy during the dry season — and the stronger the wind, the greater the rate of evaporation. Third, the clayish soil of the southern alluvial areas of Madura makes an excellent base for evaporation ponds. The finest salt came from Sumenep, where the soil had a high clay content, becoming almost rock hard with the passage of time. The salt produced in such pans was white and fine, while more porous bases resulted in coarser crystals and slightly greyish colour. Saltwater fish-farming ponds, of which there were many in Sumenep, were ideal for conversion to salt pans during the dry season (Surink, 1932: 251). Sampang also had many such fishponds.[8]

With salt no longer made anywhere else in commercial quantities, the salt industry on Madura expanded to meet the needs for the whole archipelago. People gravitated towards the centres of production to take up the work opportunities. By the standards on Madura, incomes among salt workers and plot owners were good and the Dutch emphasised these benefits for the local people. Nevertheless, any improvement in indigenous welfare was incidental to benefits for the government. There was little concern for the plight of the population of the salt-producing areas in 1869 and 1870 when production was closed down because of oversupply (van der Kemp, 1894: 267, 283; Kuntowijoyo, 1980: 185).

[8] The raising of a saltwater fish known as *bandeng* (*chanos chanos*) is a traditional livelihood along this coast. The fish grow from around one centimetre in length to a marketable size, about thirty centimetres in length, in a few months. The fish-ponds have surrounding dykes and channels leading from the sea, making them ideal for conversion for salt production. Fish farming formerly existed as an independent industry and could be carried out throughout the year, but with intensification of the salt industry under the Dutch the fish farmers were obliged to make their ponds available for salt-making during the dry season. It is common for private salt holdings to be used for fish farming during the wet season, when salt cannot be produced.

By the start of the twentieth century, the role of the Dutch in the industry had moved beyond just regulation as more and more Europeans were brought in to supervise and improve the production, processing, packaging and distribution of salt. These Dutch communities on the salt lands of Madura were quite extensive, as is evident from the long rows of Dutch houses still standing today in the rural settings of Ragung and Krampon in Sampang, and especially in Kalianget, which was virtually a Dutch town. From 1915 onwards the state salt industry was operated effectively as a state corporation under the management of its own directors. Until this time all salt holdings had been privately owned, but to achieve greater efficiency the state salt enterprise began to move into production as well, developing large holdings of its own near Sumenep and reopening on its own behalf the long-closed salt lands of Gresik, near Surabaya (Surink, 1932: 253; Kuntowijoyo, 1980: 187, 385; de Jonge, 1993: 168).

The amounts of salt produced varied according to the weather in any year, as a single shower of rain could destroy a month's work.[9] However, annual production increased steadily, reaching almost four million *pikul* (218,579 tons) in 1913. The revenues from the sale of this salt amounted to 12.6 million guilders; three years later the amount reached 16 million guilders (ENI, 1917–1927, IV: 867). Apart from the holdings of the state enterprise, the area under private salt production increased steadily. Especially was this so in Sampang, where private salt holdings almost doubled in extent between 1920 and 1933 to become by far the largest private salt production area in Madura (Kuntowijoyo, 1980: 407).

In 1930 total production was 272,000 tons, with approximately 75 per cent of that coming from private holdings and the remainder from the state salt lands (ENI, 1917–1927, IV: 867; Surink, 1932: 254). The directors of the salt industry were nevertheless dissatisfied and maintained that the industry would be more efficient if the state were to purchase all private salt lands to become the sole producer. Against intense opposition from private producers, this plan was

[9] The salt extraction process lasts three to four weeks, from the time seawater is lifted into the evaporation ponds until crystallisation. In effect, however, it only takes this time after rain or at the beginning of the season, as the different stages of evaporation take place simultaneously in separate ponds and sea water is added regularly. The frequency of 'scraping' the crystallised salt differs between the state-owned salt corporation and private salt producers.

finally realised in 1936 (de Jonge, 1993: 169). With this compulsory acquisition of all private salt holdings, the government's control over the salt industry became absolute.

After Independence, the Indonesian government inherited the salt lands of the colonial state. The state monopoly was retained, and in 1957 a state salt corporation, now known as P.T. Garam, was established. There was, however, much protest by former private salt producers against the continuation of the monopoly. In 1959 the government relented and allowed salt to be produced privately again, without restriction on location. Although the state corporation owned most of the existing salt lands and also had huge advantages of capital, technology and organisation, the private producers — representing the 'people's salt industry' (*pegaraman rakyat*), as opposed to the 'state salt industry' (*pegaraman negara*) — developed new salt holdings wherever they could,[10] amounting to 10,000 hectares of productive land in Madura alone (de Jonge, 1993: 169). The government was nevertheless biased in favour of the state enterprise and gave it increasing control over the purchase and distribution of salt after shortages occurred during the 1960s. Private producers were expected to channel their salt into the state distribution system through local cooperatives, but many of these cooperatives failed due to corruption on the part of the officials concerned (de Jonge, 1993). Consequently the private trade in salt thrived, especially from Madura to the coastal towns of East Java. However, some of the largest private producers do sell their salt through P.T. Garam's network of receiving depots and processing plants in Surabaya and Pamekasan.

The division between private and state enterprise in the Indonesian salt industry continues until the present. Some private producers have become wealthy but small salt producers in Madura are now facing difficult times due to competition from private producers in Java, especially in Gresik and Rembang and, more seriously, cheap good-quality salt imported from India and Australia. These two factors have resulted in a drastic decline in the price of salt in recent years.

[10] Salt cannot be produced on a coast open to waves because strong wave action would destroy the pond embankments (*tambak*). These limits of development were reached during the late 1990s in the Blega estuary, where one leading local salt producer tried three times to build new salt pond embankments almost at the point where the estuary meets the Madura Strait. Despite being reinforced with concrete blocks, the banks were each time smashed by waves.

Photo 2 Private salt ponds near Sampang. The small windmills are for lifting seawater from the channel in the foreground into the evaporation ponds, using a bicycle pedal mechanism attached by a pole to a bucket. The windmills are positioned to face the prevailing dry season wind. Although the system appears crude, it is remarkably efficient.

The small-scale producers have been further disadvantaged because their salt is not iodised, unlike the salt from the larger producers.

The Transport of Salt from Madura

With the concentration of the salt industry on Madura during the second half of the nineteenth century, the movement of salt became an important business. The government let contracts for the transport of salt, with sailing ships used for long-distance work. Kalianget was the main loading port for ships because it was close to the single largest salt-producing area and better situated than Pamekasan and Sampang for passages across the Java Sea and to eastern Indonesia. In 1894 the long-distance salt transport contract was awarded to the KPM, which, to reduce costs, sublet the work to the European firm of P. Landberg and Sons, which operated square-rigged sailing ships. In 1912 the state salt enterprise formed its own shipping division based at Kalianget (de Jonge, 1993: 168). For shorter routes, both within Madura and from Madura to ports along the north coast of Java and to Bali, the government invariably awarded the transport contracts to Chinese entrepreneurs, who relied upon local prahu operators to do the actual work. In 1894 there were 94 prahus involved in the transportation of salt in Sampang, 32 in Pamekasan and 90 in Sumenep (van der Kemp, 1894: 282). Although some indigenous entrepreneurs may have been wealthy enough to take up these contracts, the government deliberately favoured the Chinese by such means as placing the tender advertisements in Dutch-language newspapers only, knowing these would not be read by indigenous Madurese (Kuntowijoyo, 1980: 157).

The Chinese domination of salt transport came to an end in 1902, when the government awarded the contract for the transport of salt within Madura and East Java to the MSM, which had completed its Kalianget–Kamal railway line the previous year. The prospect of large profits from the transport of salt had in fact been a key factor in establishing the MSM. But just as the MSM could not secure the transport of cattle for itself, neither could it compete on even terms against the prahus for the transport of salt. The MSM had received preferential treatment from the government because its charges for transporting salt were much higher than had been the case with Chinese contractors relying on prahus. In 1908, however, the government decided to economise and resumed calling for tenders for the salt transport contract, with the result that the contract passed back into Chinese hands.[11]

For the short-haul transport of salt, prahus continued to be relied upon for the remainder of the colonial period, used by Chinese contractors (Kuntowijoyo, 1980: 292). There was also a considerable amount of direct transport of salt by prahu to more distant destinations along the north coast of Java. In the 1950s this increased as prahus moved into many of the shipping services left vacant by the departure of the Dutch. After deregulation of the industry in 1959, anyone who wanted to transport salt was free to do so and many

---

[11] The loss of the salt contract was a blow to the MSM, but it survived with the assistance of a government loan and even expanded its services. In addition to the railway, by 1931 the company operated a fleet of 25 buses, as well as 5 steamships, 6 motorships and 16 prahus (Kuntowijoyo, 1980: 294).

private salt producers had vessels built for themselves. In some cases, prahu owners themselves diversified into salt production.

Since the 1980s, with truck transport becoming much cheaper and the establishment of modern receiving depots in Camplong and Pamekasan, prahus have been used less and less for transporting salt. The only salt transported by prahu nowadays is from the south coast of Madura to ports on the other side of the Madura Strait, especially Pasuruan and Banyuwangi. For such routes prahus remain competitive with trucks because it is a very long way around by road. The vessels engaged in what is left of this trade are not large, nor are they in any way remarkable, for this is the poor end of the prahu transport market, occupied only by those operators who cannot afford a vessel that can compete in the transport of timber from Kalimantan. Nevertheless, without the long period when salt was a staple cargo for Madurese prahus, it is unlikely that the Madurese would have come to play as significant a role as they have in the transport of timber.

Timber

Until the twentieth century, the main items of commercial interest from Kalimantan were gold and diamonds and, to a lesser extent, high-value forest products such as camphor, dammar resin and bezoar stones rather than such a commonplace material as timber. Although Dutch officials determined early in the century that the forests of Kalimantan could be harvested for profit, little timber was imported into Java until the 1950s. The reason for this low level of interest in timber from Kalimantan was that Java still had substantial forest resources of its own, most significantly of teak (*tectona grandis*), or *jati* as this tree is known in Indonesia. This species, which does not thrive in the moist forests of Borneo or Sumatra, was regarded in Java as superior to all other species for the quality of its timber. *Kayu Kalimantan* — a catch-all phrase for timber from Kalimantan — was regarded as a poor substitute at best. Sumatran timbers likewise were generally regarded as inferior to teak.

Teak had been highly valued by the Javanese centuries before European arrival in the region. The Dutch soon learned that the timber of this tree, which grew in abundance in the central part of the island, was superior for shipbuilding as well as numerous other uses. The teak forests were then so extensive that early Dutch observers regarded them as inexhaustible, a view proved to be wrong during the seventeenth and eighteenth centuries as the more accessible stands were heavily exploited not only by the Dutch but also by Chinese and Javanese entrepreneurs. The Chinese were described as "avaricious bloodsuckers" of the forests, leasing whole inland villages from local rulers in order to plunder the adjacent forests for their shipbuilding ventures (Raffles, 1978 [1817], vol. I: 184–5). However, the Dutch and some other Europeans were almost as bad and used the same tactics to obtain materials and labour for shipbuilding (Peluso, 1988: 41).

With the establishment of the Dutch colony at Batavia, the demand for teak escalated into a boom. While shipbuilding was carried out mainly in Rembang, timber was needed in Batavia for general construction. In addition to the annual 'contingents' of teak timber that had to be supplied at fixed price to the VOC, a much larger volume of private timber shipments was sent to Batavia. Most of this came from the northeast coast of Java, especially Rembang, which in the second half of the eighteenth century had an export surplus of around 75,000 planks and 50,000 roundwood pieces. Timber — meaning teak timber — became the most important of all of northeast Java's exports and it was also by far the major cargo for the maritime entrepreneurs of the Rembang coast (Knaap, 1996: 117; Nagtegaal, 1996: 193). Teak was at that time also an important export of Madura, where the conditions were similar to the hinterland of Rembang. Even now, teak is the most common tree species in Madura, although good stems are rare outside government reserves.

As a result of this heavy exploitation, the teak stands near the coast of northeast Java became severely depleted during the eighteenth century. To obtain large logs, the woodcutters had to work further and further inland. With the hilly terrain and the lack of waterways in the forests, the transport of the logs to the coast became increasingly difficult (Nagtegaal, 1996: 195). Concerned over the declining availability of good shipbuilding timber, the VOC attempted to conserve the teak forests along the Rembang coast and in 1777 closed them to private felling. Apparently little heed was paid to this because in 1796 a VOC commission recommended a complete ban on logging in the coastal forests of Jepara and Rembang and a 50 per cent reduction in output from the inland forests of Blora and Jipang, where private entrepreneurs, especially Chinese, were obtaining their timber. Nevertheless, 'illicit' logging and private shipbuilding continued apace along the northeast coast. Although the Chinese were depicted as the principal villains in this continuing forest degradation, some Dutch were also culpable, with no less a personage than the Resident of Rembang enriching himself through involvement in large-scale private timber extraction (Peluso, 1990: 29; Knaap, 1996: 126; Boomgaard, 1991: 19).

The history of this early timber boom in Java is instructive with respect to the recent timber boom in Kalimantan. In terms of the arrogance of the state (the VOC), its inability to stop the onslaught on the forest resource, the duplicity and avarice of its senior personnel in the forest areas, the rapidity of forest degradation, and the importance of the boom for the local maritime transport industry, there are significant parallels. These arise from the inherent characteristics of timber as a commodity. First, wood is a very common material in both its raw (tree) and commodity (timber) states. Second, stands of trees of size suitable for commercial milling are found mainly in remote and sparsely populated places. Third, only simple and inexpensive technology is required to extract the raw material and convert it into a commodity, as well as to put it to end use. While these characteristics may seem obvious enough, taken together they form a unique set, quite different from the commodity characteristics of, for example, oil, aluminium or diamonds. These commodity characteristics of timber have had a significant bearing on what has happened in Kalimantan since the late 1960s, just as they did in Java in the eighteenth century.

Teak and 'Kayu Kalimantan'

The decline in teak stands did not mean that other timber species of Java became more important. Whereas teak was regarded as a valuable commodity, all other tree species — grouped together under the official category of 'junglewood' — were regarded as expendable with few controls on felling. Clearing of land for cultivation was thus mainly at the expense of junglewood forest. Much planting of teak was carried out during the nineteenth century but very little of junglewood. As a result, junglewood forest declined more rapidly than did teak forest (Boomgaard, 1988: 20; 1996: 25, 29). Eventually reserves of junglewood were allocated but these were often in steep country and established for hydrological reasons rather than to provide a timber resource (Potter, 1988: 138).

Following the practice in Java of using only teak for special buildings such as palaces and pavilions, the Dutch used virtually no other timber but teak for their own houses and furniture, as well as for commercial buildings such as tobacco sheds and sugar mills. Consequently, a certain prestige, even a mystique, began to be associated with teak. Yet, contrary to popular opinion, it is not an outstandingly strong wood. In strength it ranks below, for example, English and white oak, ironbark, jarrah and several other eucalypts, and quite a few other Indonesian timbers. It is very durable but no more so than several other Indonesian woods — indeed, several are rated slightly more durable (Perla, 2002: 30–5). The special property of teak is rather its dimensional stability: its resistance to shrinking and warping, even when unseasoned. It is above all for this reason, coupled with its durability, that teak has been so sought after for marine work, especially for decks and planking *above* the waterline. For the same reason it is ideal for exterior joinery items such as window surrounds and doors. In the hot and humid conditions of Java, the use of inferior timbers for such work soon results in opening up at joints, jamming in place during wet weather and rotting. While low-demand timbers could serve satisfactorily for internal structural purposes, bias in favour of teak, reinforced by the example of the European masters, was too strong to allow such an objective approach. For those who could afford it or who could take it directly, teak became throughout the late colonial period increasingly the timber of choice for both construction and joinery, while junglewood timbers — a category including most timber from the Outer Islands — remained in correspondingly low demand.[12]

For all these reasons, there was little commercial timber felling outside Java and Madura. The main exception was coastal Sumatra, where Chinese-owned companies logged large amounts for export to Singapore or for use in the tin mines of Bangka and Beliton. At the start of the twentieth century, numerous logging concessions in Sumatra were applied for and granted, but most never reached the production stage because of technical and logistical problems. As late as 1925 there was still little competition for the leading Chinese

---

[12] Some junglewood timbers of Java could have been more widely used for exterior construction. One example is *memba* (*azadirachta indica*), which grows fairly prolifically. In recent years this timber has been much used in house construction, with good results. It has also been found to be satisfactory for marine use, and is now often used in parts of the south coast of Madura for the building of fishing craft. In Bali, where it is called *intaran*, it is similarly now recognised as a first-class construction timber.

timber merchants, who were then responsible for two-thirds of all timber being cut in Sumatra (Boomgaard, 1996: 30). It is thus hardly surprising that the vast forest resources of Kalimantan were still not being commercially exploited because, in addition to the poor reputation of most Bornean woods for durability, the logistical difficulties were if anything worse than in Sumatra.

Nevertheless, there was a demand for a few dense and extremely durable timbers from Sumatra and Kalimantan. The most important of these was ironwood (*eusideroxylon zwageri*), known in Indonesia as *ulin*.[13] This timber, moderately abundant in eastern Borneo, and at one time also in South Sumatra,[14] is considerably stronger than teak and more durable as well. It also has good dimensional stability and is one of the most resistant of all timbers to attack from termites and marine borers (Soewanda *et al.*, 1971: 55; Chudnoff, 1980: 643; Perla, 2002: 35). In colour it is a deep reddish brown, weathering to dark grey, and is available even now in South Kalimantan in long clean baulks up to 20 metres in length. It is the undisputed 'king' of Bornean timbers, the traditional timber of choice not only for primary construction timbers such as longhouse posts, but also for roofing shingles.[15] Ulin was also used extensively by the Dutch for construction of their oil installations and coal mines, as well as for railway sleepers. It is also an outstanding marine timber and suitable for any part of hull construction, including the dowels used for edge fastening of planks, for which purpose teak is not regarded as strong enough.[16]

It was mainly the profit potential of ironwood which underlay a rash of applications by Dutch companies for logging concessions

---

[13] The term 'ironwood' has also been widely used in other parts of the world for dense hardwoods not related to *eusideroxylon zwageri*. For this reason the indigenous name of *ulin* will be used here, except when referring to works in which the term 'ironwood' was used.

[14] The formerly extensive stands of *eusideroxylon zwageri* in South Sumatra have been almost entirely destroyed (Collins, Sayer and Whitmore, 1991: 147).

[15] Roofing shingles in Kalimantan are made from pieces about 10 mm thick by about 110 mm wide. They are not sawn to this thickness, but split from short blocks of wood. Despite its hardness, ulin cleaves well due to its lack of interlocking grain.

[16] In Madura, where teak is the traditional timber for hull planking, the dowels securing the planks together are usually made from the wood of a small local tree called *pereng*.

in Borneo early in the twentieth century. However, the conditions proved even tougher than in Sumatra and all concessions granted up to 1916 failed. Although many ventures were undercapitalised and probably speculative, even groups with substantial capital soon folded. One well-capitalised venture was the Surabaya-based Seliman Hout-en Landexploitatie Maatschappij (Seliman Timber and Land Exploitation Company), which in 1913 set up a modern sawmill at Berau on the east coast of Borneo with a staff of 200 Javanese workers and its own steamship for transporting the timber to Surabaya, but by 1915 was on the verge of liquidation (Potter, 1988: 135; Lindblad, 1988: 102). It seems that all of these concerns underestimated the magnitude of the difficulties involved in the extraction of timber from the forest. A problem with ulin, along with a number of other particularly desirable species, is that it does not float, and thus cannot be 'rafted' downriver without buoyancy support.[17]

Nevertheless, the lure of profits ensured steady applications for further concessions, while the government was anxious to see some positive return from this vast resource. From 1916 there was a strong temporary demand in Java for construction timber, with thousands of houses being rebuilt under a housing improvement scheme motivated by concern over declining native welfare, while ulin was also increasingly being used as an alternative to teak for railway sleepers. Yet again success proved elusive: of 44 forest concessions granted in Southeast Borneo between 1921 and 1924, none were active by the end of that decade. Little of the timber that was obtained was felled by the concession holders, who instead found it easier to rely on local Dayak workers, paying them for each tree felled on a casual procurement basis. This proved to be a poor system, with the logs remaining

[17] Ulin weighs over 80 lb per cubic foot when freshly cut and between 64 and 71 lb per cubic foot fully seasoned (Chudnoff, 1980: 643). Logs to be transported from the forest would almost never be fully seasoned anyway, as this would require several years for such a large piece. The statement by Potter (1988: 129) about "Borneo ironwood" being "floated down the Barito River" is thus misleading. The most likely means of 'floating' would have been by supporting the logs with a pair of large river sampan, one to each side of the log or logs, and lashed together. Where logs have been removed from an area by floating them downstream, it has in fact been common practice to avoid taking 'sinker' species (Collins, Sayer, and Whitmore, 1991: 46).

for such a long time on the forest floor or in the rivers that by the time they were finally delivered many of them were badly affected by borers and rot, further eroding the already dubious reputation of most Bornean timbers by comparison with teak (Lindblad, 1988: 102–3; Potter, 1988: 137).[18] Thus from an estimated 275,000 cubic metres of timber cut in South and East Borneo in 1925, only 44,523 cubic metres were exported, with about 60 per cent of that going to Java and the remainder to East Asian destinations. The main consumers of timber were the coal and oil industries, which also bought timber from Dayak fellers despite having their own concessions. The difficulties of breaking away from the system of buying logs from Dayak cutters were made clear in the same year with the failure of a venture that had been contracted to supply ironwood for railway sleepers to Java without purchasing from Dayak cutters. Apart from the problems of logistics and capitalisation, the market in Java was unreliable for anything else than ulin, with frequent cancellations of contracts from buyers. Probably for this reason, timber entrepreneurs in the major east coast ports of Balikpapan and Samarinda preferred to sell to buyers from East Asia rather than Java (Potter, 1988: 138).

The export trade remained small during the early stage of the Depression, with prices so low that logs were simply left on the ground, and many of those that did arrive downriver were unacceptable to the foreign market. Then in 1932 a large Japanese venture commenced operations near Tarakan, employing over 1,000 workers despite not even having a logging concession. Huge amounts of timber were bought from the local population and shipped off to Japan without any customs formalities (Potter, 1988: 141; Lindblad, 1988: 104). This venture was subsequently forced by the Dutch to move to Sangkulirang, to the north of Samarinda, where it could be monitored, but the activities of the Japanese had already forced a significant change in the timber regulations. Under the Timber Ordinance of 1934, all 'wild cutting' was prohibited, so that concession holders would thereafter have to fell their trees themselves instead of buying logs from Dayak fellers. Indigenous people would be allowed to cut timber only for their own consumption.

[18] These logs can be presumed not to have been ulin, which is highly resistant to borers as well as being almost rot-proof, and in any case too heavy to float.

The Japanese concern at Sangkulirang became even bigger, employing at one stage 3,500 labourers under 200 Japanese supervisors. Although labour officials were highly critical of the working and living conditions in the camp and the government also had misgivings about the long-term intentions of the Japanese, these misgivings were tempered by a sense of relief that this huge forest resource was finally starting to yield some economic benefit. There were also other, smaller Japanese companies involved in timber extraction on the east coast, while a Dutch firm was operating a profitable sawmill on the island of Nunukan, near Tarakan on the northeast coast. All these big logging concerns in eastern Borneo were foreign enclaves and they all used their own or chartered ships to export their timber. Even for the trade within the archipelago, prahus were not involved. In the case of the Nunukan venture, the Dutch company used its own steamship to transport ironwood to Tanjung Priok (Potter, 1988: 142; Lindblad, 1988: 105).

Despite the success of these ventures, timber output from Dutch Borneo remained modest by comparison with that from Java and Sumatra.[19] However, the government envisioned further expansion of the industry in Borneo, especially in the south and southeast, where exploitation had been light. In 1941 plans were made for state-managed forest exploitation based at Sampit, on the Mentaya river on the southern rim of Borneo, and at Batulicin on the southeast coast opposite the island known as Pulau Laut. The Sampit operation intended to concentrate on the harvesting of a softwood species (*agathis alba*) for the manufacture of plywood, with the processing taking place in Java, while at Batulicin the species targeted was to be ulin, which was plentiful in that area. However, the war and the subsequent revolution disrupted these plans as well as those for further exploitation in the forests of the northeast. In 1949 production in the northeast remained low, at less than 30,000 cubic metres, or perhaps a tenth of the amount achieved in the immediate pre-war years. In the south output was better, at about 90,000 cubic metres for 1949,

[19] In 1941, on the eve of the Pacific War, total timber production from Borneo reached 341,000 cubic metres, with most of that coming from the northeast concessions. This compared with 796,000 cubic metres for Sumatra and 730,000 cubic metres — over 90 per cent teak — for Java (Boomgaard, 1996: 149, 168).

with around half of that coming from the government-owned plywood company based at Sampit. As before the war, prahus still had no significant role in the transport of timber, with the Sampit company using two small motorships of its own to carry timber to Java for processing. This concern continued to operate after the departure of the Dutch, first under the auspices of the Indonesian government until 1962 and thereafter under the management of the state-owned timber corporation Perum Perhutani (Potter, 1988: 143–4; Boomgaard, 1996: 168).

Exploitation for Foreign Exchange under the New Order

Before 1966 little timber was exported from Indonesia because it was easier and cheaper for foreign timber companies to obtain their supplies from the Philippines or Sabah. Kalimantan had three significant disadvantages. First, the political climate in Indonesia in the early 1960s and the progressive nationalisation of foreign companies discouraged foreign investment. Second, infrastructure was almost non-existent with few good roads, requiring large initial outlays on the part of logging companies. Third, the Basic Agrarian Law of 1960 acknowledged the usufructory right of traditional forest-dwelling peoples to take what they needed from the forest, which in effect sanctioned shifting cultivation on forest land and complicated the matter of concessions.

In 1967 the New Order government, having inherited a disastrous national economy and being aware of the wealth that had flowed into neighbouring countries as a result of forest exploitation, resolved to attract foreign investment into the timber industry. To this end it offered very generous tax concessions to foreign timber companies, lowered the royalties payable in order to undercut competition from the Philippines, and gave a clear assurance in law that indigenous usufructory rights would not stand in the way of commercial timber extraction. These measures had the desired effect. Massive foreign investment poured into the forestry sector, mostly in the province with the richest forest resource, East Kalimantan. Earnings from timber exports soared, from US$4 million in 1966 to US$3,200 million in 1973 (Ross, 2001:169). Most of this timber was in log form, making Indonesia the world's biggest log exporter.

It was not big companies that were initially responsible for this huge upsurge in production. Until 1970 only a few foreign companies had managed to obtain concessions because the procedures for obtaining large concessions remained so complicated. On the other hand, many small concessions were given by regional governments to local applicants. In practice, concessions were often given to political favourites with local businessmen and politicians working together (Magenda, 1991: 78–85). For the first three years of the boom, it was these small concession holders who dominated log production in East Kalimantan. The logs were cut and hauled manually and floated downriver during the rainy season in a 'log flood' when the rivers were high, to be sold to foreign buyers for prices well beyond the level of the domestic market. It was a bonanza time for local entrepreneurs large and small and almost all the able-bodied men in forest areas were involved in some degree in this trade (Peluso, 1983: 179).

Most of the logs exported through this 'log flood' system were lighter hardwoods, mainly dipterocarp species. Not only were these dipterocarps the most abundant of the larger tree varieties but, with a few exceptions, dipterocarp logs also would float even when freshly felled. Although these lighter hardwoods were less strong and durable than ulin and some other species suitable for exterior construction work, this was of little importance for the purchasing companies because much of the timber from Kalimantan was to be used in the plywood industry.

Since ancient times woodworkers had tried to overcome the principal structural shortcoming of timber: its weakness across the grain and the consequential difficulty of making wide panels. Before World War II most of these attempts had been too expensive, laborious, or lacking in durability to be of much use. The forced wartime development of modern glues, rotating veneer cutters and press machinery finally overcame this problem. The resulting material — thin layers of wood glued together with the grain in each layer at right angles to that in the adjacent layer(s) — was stronger for its weight than steel and enabled wood to at last be used in large, thin panels.

Durability had always been one of the most sought-after qualities of timber, along with straightness and absence of flaws. Cheap non-durable timber had its place but its inferiority was always apparent, nowhere more so than in the humid tropics. For plywood, however, it was the lighter and less durable hardwood species that were preferred, not just because they were easier to obtain and process but also because their veneers were relatively stable, evenly grained and free from knots and other flaws. These qualities were most obvious when compared to veneers from many softwood species, especially of the fir group. Apart from the physical characteristics of the wood, these logs from Kalimantan also had the great advantage of being cheap. The more durable construction timbers remained important for sawn timber but it was the formerly overlooked species, with only moderately durable timber despite the size of stems and relative absence of flaws, that now came to dominate in the tropical timber industry.

A fundamental shift in timber use and demand had thus taken place in Kalimantan, with Japanese interests playing a leading role. Japan had become the world's leading producer of tropical plywood during the 1950s, relying on logs obtained from the Philippines. Much of its output had been exported to the USA but the Japanese market soon began to absorb vast amounts of this new panel-wood material for its own needs. Apart from conventional panelling and partitioning, huge amounts of plywood were used in Japan for concrete formwork, for which durability and high quality hardly mattered because the panels were only used two or three times before being discarded (Dauvergne, 1997: 8, 176, 182).

Concentration of Business Interests

Small-scale involvement in the export timber trade was effectively stopped in 1970 by a presidential decree that stipulated a minimum size of 50,000 hectares for forest concessions and also that all logging operations be mechanised. The latter requirement was supposedly to comply with Japanese demands for logs in better condition than apparently obtained under the 'log flood' system. Most significantly, the new law also transferred the issuing of forest concessions from regional governments to Jakarta. This made it much easier for foreign companies to obtain concessions, but they were now required to take on Indonesian partners. In the event, many of these partner firms had close connections with the military, as well as with political and bureaucratic elites (Gillis, 1988: 74; Ross, 2001: 176). Political patronage was probably the main driving force behind these changes.

Although profits continued to be huge, with minimal outlay and no effort on the part of military-backed firms, timber rents achieved by the state remained low. Meanwhile, silvicultural obligations, which had been introduced in 1972 in a sincere but probably naive attempt to make the industry sustainable, were rarely complied with (Gillis, 1988: 84; Dauvergne, 1997: 75; Ross, 2001: 179).

In 1981 the government moved to phase out the export of logs and force the development of secondary industries, especially the manufacturing of plywood. After 1985 there was a total ban on the export of logs from Indonesia. Timber companies with no downstream processing had to invest in that direction or sell their logs cheaply to other firms with plywood factories. Many local timber companies went bankrupt. The foreign timber companies withdrew and their military-backed partners faded away, the industry becoming dominated instead by a small number of Sino-Indonesian concerns, some with close ties to Japanese firms (Dauvergne, 1997: 87). From an army of small entrepreneurs in the 'log flood' period, the industry was now highly concentrated in elite hands, with some of the largest concession holders having close links to President Suharto. Indonesia now became the world's largest producer of tropical plywood. In theory this concentration on wood processing and value-adding should have had some conservation benefits because investors in sawmills want to be assured of log supply. In practice, the industry became overcapitalised, with the proliferation of plywood mills placing even heavier demands on the forest resource than at the height of the log export boom in the late 1970s.

The Demand for Sawn Timber in Java

The timber trade from the Outer Islands to Java developed roughly in parallel to the foreign export trade but with little direct connection to it. There was not even much overlap in the supply areas for the two trades because the export trade took place mainly from East Kalimantan, whereas the Java-bound trade was supplied mainly from Banjarmasin in South Kalimantan and the southern Sumatran provinces of South Sumatra and Lampung. Although some timber for the domestic trade was carried by motorships, from early in the twentieth century prahus had been the main carriers in the domestic trade. By then steam shipping had marginalised the prahu fleets and relegated them mainly to a feeder role, but after the end of World War I prahus began to appear more often in Surabaya with loads of timber from Banjarmasin (Dick, 1987: 106). Timber was not a popular cargo for regular motorised shipping services because loading was slow and the low freight rates gave a poor return.

The Banjarmasin–Surabaya route was ideal for prahus because during the southeast monsoon they could usually make the passage each way across the Java Sea without tacking. In 1922 the total amount of cargo exported by prahu from Banjarmasin was about 14,000 *last*, this presumably going to Surabaya and the *Oosthoek* (East End) ports of Pasuruan, Probolinggo and Panarukan, which were used mainly by Madurese prahus. Probably most of this was timber, for the only other significant export good from Banjarmasin was woven mats, used mainly for packing in the sugar and salt industries (Dick, 1987: 106). Nevertheless, this flow of timber from Banjarmasin to East Java was insignificant compared to the production of teak in Java, with an average of 218,000 cubic metres of sawn teak produced for each of the years 1920 to 1925 (Boomgaard, 1996: 149).

The KPM regarded the presence of prahus on the Banjarmasin–Surabaya route as 'trespassing' on its own domain. It responded by slashing freight rates for timber and woven mats. The prahu trade between Banjarmasin and Surabaya shrank so much that by the mid-1920s only the predominantly Madurese Oosthoek ports were significant destinations for prahus from Banjarmasin (Dick, 1987: 106). The timber then exported from Banjarmasin would have been mainly dipterocarps, especially *meranti* (*shorea* sp.), the predominant commercial species of the Barito river system (Potter, 1988: 135). Unlike on the east coast, ulin is not common in southern Borneo west of the Meratus range. There are several different varieties of meranti, falling into the market categories of white, light red and dark red, with considerable differences in density.[20] Other common commercial timber categories from the Barito-Kapuas area (and elsewhere in Borneo) are *keruing* (*dipterocarpus* sp.), a pale-coloured timber of moderate strength and durability but appreciated for its consistency and grain, and *kapur* (*dryobalanops* sp.), of similar strength and durability but with less consistency and coarser grain. However, none of these common timbers can compare with teak for durability and dimensional stability. On the market in Java, they were regarded just as alternatives to the local junglewoods.

By the mid-1950s Java's teak forests, which had been carefully managed by the Dutch, were much degraded. The decline had started during the Japanese Occupation, when the Japanese indiscriminately cut wood for construction projects and as fuel for trains and factories. At the same time local villagers, freed from what they had seen as the tyranny of the Forest Service, cleared much forest land for agriculture.

---

[20] In the United States meranti is marketed under the Philippine name of *lauan*. It is sometimes also misleadingly referred to as 'Philippine mahogany'.

During the three and a half years of the occupation, timber and fuelwood consumption in Java was approximately double what it had been under Dutch management, while very little planting of trees was undertaken. Rampant destruction continued during the revolutionary years, with much deliberate burning of forest, occupation by squatters, and looting of timber (Peluso, 1990: 40–3).

Steep increases in the price of both teak and local junglewood inevitably made the prospect of timber from South Kalimantan more attractive than it had been before the war. When a construction boom occurred in the mid-1950s, Bugis and Makassarese prahu operators based in Surabaya and Jakarta were able to take advantage of the strong demand for timber by supplying much cheaper stock from Kalimantan and Sumatra (Dick, 1975a: 91). However, not until the economy and construction began to surge ahead under the New Order did timber from Kalimantan become broadly acceptable on the Java market and come to be seen as the solution to Java's forestry problems. Although caution is needed in interpreting port statistics on timber traffic, figures from the early 1970s show that the domestic timber trade was booming. Outward cargo from Banjarmasin, the most important centre for domestic timber supply, more than doubled between 1967 and 1972 (Dick, 1975a: 83). Similarly, timber imports for Java and Bali doubled between 1971 and 1973, with the volume imported in 1973 reaching about 700,000 cubic metres (Kehutanan, 1974, B: 6). Jakarta was by far the most important receiving port, followed by Surabaya (Dick, 1975a: 94).

The timber now flooding into Java was being used mainly in the construction industry, especially for housing. This was despite the fact that most modern houses in both urban and rural areas of Java and Madura are constructed with brick walls on stone foundations, with the brickwork usually plastered over. Floors are generally of plaster or concrete render and, in more affluent households, covered with ceramic tiles. Unlike in Malaysia or Thailand, parquetry flooring is rare. Yet despite this reliance on brick and concrete,[21] much timber is still used in house building, for framing and doors.

[21] In the early 1970s it was estimated that annual per capita wood consumption in West Java was 0.08 cubic meters. This compared with 0.24 cubic metres in Malaysia, 0.5 cubic metres in Taiwan and approximately 1.0 cubic metre in Japan (Kehutanan, 1974, A: 2–3).

Transportation

Most of the timber coming into Java in the early 1970s was being carried by prahus. The domination by prahus of this trade was regarded by senior forestry department officials as unsatisfactory. The great majority of prahus were still engineless and were regarded as slow and unreliable, while their destination ports, especially Sunda Kelapa in Jakarta, were regularly clogged up with many vessels waiting their turn to berth. These senior officials wanted an efficient modern timber transportation service. In the case of ports with a high potential for timber export and which could be entered by ships, such as Banjarmasin in South Kalimantan and Sampit in Central Kalimantan, it was hoped to use specialist log-carriers, or at least regular liner shipping services, for transporting timber to Java. Where ports were too shallow to be entered by ships, the preferred method of conveyance was by barges towed by tug vessels. Prahus were regarded as acceptable only for the shipping of timber from small ports (Kehutanan, 1974, B: 11, 20).

This expectation that the carriage of timber should be mainly by modern shipping ignored the fact that prahus were already meeting the demand for sawn timber in Java, as reflected in the steep increase in the volume being landed. Indeed, the Jakarta market was over-supplied at times, with meranti being 'dumped' at discount prices (Kehutanan, 1974, D: 1). Only in the transport of logs, as required by plywood factories, were prahus inadequate. However, that would not have mattered if plywood factories were located in or near supply areas, as became the case after the swing in the late 1970s towards large-scale domestic processing.

Besides meeting the domestic demand for imported timber, prahus offered much cheaper freight rates than modern shipping. Given the concern in forestry department circles over the price of timber from Kalimantan and the implications for the environment in Java, the performance of prahus as the principal carriers of timber should have been welcomed by the government. The negative attitude of these bureaucrats towards prahu shipping reflected the view that as a newly industrialising country, Indonesia should not have to depend on a third-world maritime transport service, regardless of how economical or efficient that service might be. After all, the timber exported from Kalimantan to foreign destinations was being carried by specialist log-carriers and during the 1930s the Dutch had been using steam

Photo 3 Unloading timber by hand at Kali Baru, Jakarta.

shipping for the transport of timber from the east coast of Borneo to Jakarta. The continuing domination of the domestic supply trade by prahus was thus perceived as an embarrassment in some circles of the New Order state.

Beyond their failure to appreciate the capability and economic advantages that prahus could offer, senior forestry officials also showed a poor awareness of the special problems of timber as a transport commodity. Because of its bulk and awkwardness in handling, timber has always been a low-paying cargo, normally becoming attractive only as a return cargo after a more lucrative outward voyage (Dick, 1974a: 95). Apart from the poor freight rates, in Indonesia timber is also a slow cargo to load and unload. Planks are carried one or two at a time by individual labourers traversing precarious gangplanks between vessel and quay, in marked contrast to rapid container handling or mechanised bulk loading and discharge. If timber can be obtained cheaply and be sold elsewhere for a good price, as the foreign companies exploiting Southeast Asian forest resources from the 1950s onwards were able to do, the higher costs of modern shipping transport can be absorbed. However, no such advantage applied in the Indonesian domestic market, where the wage structure at the destination was even lower than that prevailing in the loading ports of Kalimantan.

Indeed, it seems that from the start of the timber boom, forestry department bureaucrats had been out of touch with the reality of the domestic timber trade. In 1974, by which time timber from the Outer Islands had for several years been coming into Java on a large scale, the Directorate of Timber Marketing in Jakarta stated that "so far, we know little about the production of timber [for the domestic market], the movement of it, and what happens to it once it has reached consumer areas" (Kehutanan, 1974, B: 14). This was a remarkable admission, coming as it did from a government agency that was supposed to be coordinating, or at least monitoring, the domestic timber trade.

The main reason for this lack of knowledge on the part of these bureaucrats about the nature of the domestic timber trade was that the government's main priority was the vastly more profitable foreign export trade. Not only was timber by 1974 a big earner of foreign exchange, ranking second only to oil, but many senior figures in the New Order government and the military were also profiting personally from the foreign timber trade. They were able to enjoy the economic rents from the timber industry through the offices of Suharto's trusted forestry minister, Sudjarwo, who arranged lucrative deals for presidential clients as well as for Suharto's family members (Robison, 1986; Aditjondro, 1998; Ross, 2001). By comparison, there was little interest among senior New Order figures in the domestic timber trade. Similarly, neither the foreign timber companies nor their Indonesian partners had any interest in supplying the low-priced local market.

Although the domestic trade boomed more or less at the same time as the foreign export trade, it was not directly connected with the latter and developed simply as a result of local market forces, without any government intervention. However, there was an indirect connection between the New Order's export timber trade and the domestic trade. The onset of large-scale logging in Kalimantan by foreign companies sent a clear signal to the Banjarese, the ethnic group which has been appropriately called the 'homo economicus' of southern Kalimantan (Lindblad, 1988: 59), that the time was ripe for large-scale timber extraction for the domestic market. Likewise, in Sumatra and West Kalimantan, local entrepreneurs recognised the opportunities in the Java market and expanded their operations accordingly. Despite being conducted on a large scale since the late 1960s, this was a fringe economy with very little of the timber landed in Java coming from the government's forestry concession scheme (*hak pengusahaan hutan*, HPH). Nearly all big concessions were instead supplying the foreign export market.

The critical link in the domestic trade, and the riskiest, was the transport from places of origin to Java. Whereas timber dealers anywhere could make a profit by selling timber for a price well above that which it had cost them, to profit from the movement of timber from Sumatra and Kalimantan to Java was more difficult. With the low freight rates and the long times spent in port at either end, this was a niche that could be occupied only by informal-sector entrepreneurs operating on minimal budgets. The Madurese, masters of the niche economy, were well qualified to take advantage of this opportunity. As will be shown in Chapter 4, however, their approach to business in the timber trade inevitably brought them into conflict with the state.

3

Madurese Prahus and Their Home Coasts

Until the modernisation of Indonesia's prahu fleets from about 1980 onwards, the prahus of the Madurese were the most distinctive and traditional of all the types operated by the leading maritime ethnic groups. They were also the most diverse, with three main types in use almost throughout the twentieth century, whereas the Bugis, Mandar and Butonese fleets were each dominated by a single generic type. For the modern observer, the anachronistic and even archaic appearance of these Madurese prahus evoked a sense of a maritime tradition going back centuries. Nevertheless, appearances were in this respect deceptive because these prahus were quite different from those used by the Madurese in the nineteenth century and earlier.

While this chapter is principally concerned with these Madurese sailing prahus of modern vintage, it opens with an overview of the main types used by the Madurese before the twentieth century. Apart from nautical historical interest, this overview of earlier vessel types is instructive in what it reveals about the Madurese themselves and their relations with other maritime peoples of the region. After this scene setting, the greater part of the chapter focuses on the dominant twentieth-century types, with particular attention to their geographic distribution around Madura and the economic niches occupied by local fleets. The descriptions of the vessels are general with the emphasis on their character rather than construction details; readers interested in technical aspects and their lineages and development are referred to Appendix 1. The last part of the chapter looks briefly at the types of vessel currently used by the Madurese, together with notes on the regulations governing prahus and the economic significance of these regulations for prahu owners.

Figure 1 A Madurese vessel used for ship-to-shore work in the Surabaya roads in the 1830s, from a sketch by Paris (1842: Plate 89). Although Paris described this vessel simply as "a transport boat", the construction, ornamental features and sailing rig (here not in use) all point to it being Madurese. Such boats were primarily powered by oars, which is why they were steered by a sweep rather than a true lateral rudder. The single triangular sail was used only when the wind was from a fair direction. In this example, the sail has been furled and stowed together with its spars on the deckhouse roof to reduce wind resistance.

Early Modern Madurese Vessels

The Madurese have long been known for their boats. In the early fourteenth century, a Chinese author, Wang Ta-yuan, wrote of the Madurese:

> They make boats of wooden boards and fasten them with split rattan.... The hull is very flexible, and rides up and down on the waves, and they row them with oars made of wood, too. None of them has ever been known to break up (Cited in Scott, 1982: 340).

Two centuries later, the Portuguese chronicler Pires, who did not visit Madura but heard plenty about it in nearby Tuban and Gresik, noted that it was a place with "many lancharas" (Cortesao, 1967 [1944]: 227). The use of the term 'lanchara', which is derived from the Malay word *lancar*, meaning 'swift' (Mills, 1930: 159), suggests vessels that were light and fast rather than burdensome. Such vessels were probably undecked boats of between eight and fifteen metres in length, propelled by both oars and sails. This interpretation fits with the earlier Chinese description and also with the illustration of a Madurese boat, although not described as such, in Paris (1842) (see Figure 1).

The association of Madura with vessels of this sort continued into the modern era. In 1903, of the more than 8,000 fishing vessels in Madura, 1,720 were *mayang*, moderately large open vessels crewed by 14 or more men and propelled by both sail and oars (OMW, 1905: 18).[1] Nevertheless, the *mayang* is better known as a Javanese type. Indeed, it is the quintessential Javanese vessel type, with large fleets scattered along the entire north coast of Java until almost the end of the twentieth century. Although the name of the type indicates a fishing vessel, the *mayang* was formerly also widely used for cargo work. Indeed, in the late eighteenth century it was the commonest of all transport prahu types along the north coast of Java (Knaap, 1996: 38). Many were given deckhouses to make them more suitable for this purpose. So modified they were not strictly *mayang* because the deckhouse prevented the use of the *payang* net from which the name was derived, but the nomenclature stuck.

The *mayang* was also the most common transport prahu in Madura in the late eighteenth century (Knaap, 1936: 38, 66). As noted above, the type was still common in Madura as a fishing vessel in the twentieth century. This widespread use of apparently the same vessel type conveys the impression of a common maritime heritage for Java and Madura. However, there is strong reason to believe that the *mayang*, a vessel of highly distinctive form and construction (Burningham and Stenross, 1994), was not autochthonous to Madura but introduced from Java.

Whereas the *mayang* was shallow and broad-beamed, the quintessential vessel of Madura was of a completely different sort: a double-outrigger craft known as the *peduang* (*bedouang*, *paduwang*). There were two main types of *peduang*. While the most common had a dugout canoe hull and was used mainly in fishing, the largest version was highly developed, being fully planked with a rather boat-like form,

---

[1] The name *mayang* means a vessel that uses the *payang* (umbrella) net, a traditional type of purse seine net. These nets and their associated gear were large (see Surink, 1926), as they still are today, necessitating a large crew and boat to carry them. But there were upper limits because the boat needed to be rowed a lot during the fishing. Since about 1980 the oars have been replaced by side-mounted engines, with the net worked on the opposite side to the engine.

Photo 4 A Madurese *mayang* at Besuki, on the mainland of East Java. This type is called a *jangraja*, with reference to the exaggerated bow structure (from *pajang*, 'display' and *rajo*, 'big').

Photo 5 Another *mayang* at Besuki, viewed from ahead to show the shallow multi-chined form typical of boats of the *mayang* family. The object projecting ahead of the bow is a wooden anchor.

Figure 2 A *peduang* sailing off Besuki in the 1830s, from a sketch by Paris (1842: Plate 92).

a low deckhouse and massive outrigger beams. These large *peduang* were 14 to 16 metres in length with a capacity of several tons. In the opinion of the maritime ethnologist James Hornell, who saw some in the early twentieth century, they were "by far the finest outrigger coasting [trading] craft in the Dutch Indies" (Hornell, 1920: 99).

Peduang were not much used for long-distance work, although they were capable craft at sea, but they were very common in the Madura Strait until the end of the nineteenth century. The decline of the type was not just because it was less capacious than other types of prahu but also because the outriggers were unsuitable for the increasingly crowded prahu harbours of Java. For this reason, by far the most important port of call in Java for *peduang* was Besuki, which is still an open beach 'port'.

The primacy of the *peduang* as the autochthonous vessel type of Madura has been historically blurred by the dominance of Javanese vessel types for trading purposes. A clearer view is obtained from the large island of Sapudi, one of the cradles of Madurese culture and an important place for seafaring. Being situated well to the east of the main island of Madura, it was further removed from the influence of Java. In 1890, the entire fishing fleet of 228 vessels at Sapudi consisted of *peduang*, while for trading vessels 31 out of a total of 70 vessels were also *peduang*, the remainder consisting of a mixture of non-outrigger types (Verwijk, 1893: 237–44).

Meanwhile, the Madurese had for centuries also been building their own small- and medium-sized utility boats, of quite different construction from the classic Javanese type, the mayang. As vessel builders, however, they were less versatile than the Javanese craftsmen of the Rembang coast, who also had the advantage of better teak supplies. For these reasons many Madurese entrepreneurs of sufficient means preferred to purchase vessels for trading work from Java. This trend is illustrated by another important prahu type, the *pencalang*.[2] This was originally a Malay type, dating back at least to the seventeenth century, which gradually found favour along the entire north coast of Java (Knaap, 1996: 34, 155). The *pencalang* was also much used by the Madurese. In the 1770s it was the third most common trading vessel type associated with Madurese skippers, while in 1780 there were eight *pencalang* in the private fleet of Cakraningrat, the ruler of West Madura. Most examples of the type were about the same length and beam as large *mayang* but with greater depth of hull giving increased capacity and seaworthiness, a spacious deckhouse, and a raised cabin aft for the crew.[3] It was a true little ship and the vessel of choice for trading purposes, especially over long distances. A handsome and well-proportioned vessel, it was often embellished with ornamental features. It was also a good sailer, with leaner versions much used by pirates.

In the early twentieth century, *pencalang* were still very popular on the south coast of East Madura. They were referred to in reports as *cemplon*, the name used for the type in northeast Java since at least the mid-nineteenth century (for example, Hageman, 1862: 288) and

---

[2] Pronounced with a slight stress on the second syllable.

[3] It was not however "a large mayang with a deckhouse" as asserted by Horridge (1981: 50). While many *mayang* were so modified (see, for example, Horridge, 1981: 47; also 1986: 11), such vessels were significantly different from the *pencalang*. An illustration of a *pencalang* appears in Paris (1842: plate 86), although there it is incorrectly described as a "prao mayang". Dutch officials, and some other observers, were often less than meticulous about the names they gave to indigenous vessel types. The men who built and worked on the vessel illustrated by Paris would probably have called it a *cemplon*, the standard name for the type in East Java. Another illustration of the same type, from Semarang in 1898, appears in Nieuwenkamp (1926: 148–9). The accompanying caption describes it simply as "a Buginese vessel", which is perhaps not incorrect because *pencalang* had long been popular with sailors from South Sulawesi (see Knaap and Sutherland, 2004: 65).

Photo 6 The last *kacik*, named *Lanceng Tresno* (Lad in Love), in the Madura Strait, 1984.

still in use in the 1980s. However, the Madurese who used these vessels called them *kacik*. In many cases, especially in Pamekasan, prahus of this type were purchased second-hand from Java, probably from Rembang or yards near Tuban. Such prahus were usually old and in poor condition and had to be repaired or extensively rebuilt for use in Madura (OMW, 1906: 18). Some *kacik* were also built in Madura and to a high standard, although probably few Madurese possessed the skill needed for this. The type gradually lost favour because it was difficult and expensive to build, with extensive ornamental carving work an integral element of the design. The last *kacik* was in use until as late as 1984, operating a twice-weekly service between Talang, on the south coast of Madura not far from the town of Pamekasan, and Besuki on the other side of the Madura Strait, carrying petty traders and villagers going to market just as had been the case for at least the two previous centuries (Photos 6, 7).

The last two vessels discussed above, the *peduang* and the *kacik*, the Madurese version of the *pencalang* or *cemplon*, represented the two extremes of Madurese trading vessel types: the purely indigenous outrigger coaster and the little ship from the north coast of Java, with

Photo 7 The stern of *Lanceng Tresno*, showing the elaborate ornamentation typical of the *kacik*. After it ceased service, this vessel was left derelict in a muddy reach near Talang, disappearing altogether after a few years. The ornamental panel above the stern was saved, however, and in 2000 was still in Talang, in the home of the late owner's sister.

most other vessel types coming somewhere in between. However, there was another important group of prahus associated with the Madurese that lay outside this broad range. The prahus in this further group were limited to Sumenep and consisted of types from South Sulawesi. These had either been introduced by Bugis and Makassarese migrants who had settled in the outer islands of Sumenep, or had been purchased directly from Makassar. There were four types of prahu in this group, all developed from the same basic type: the *pajala*, the standard vessel of Makassar, usually undecked; the *sope* or *sopet*, as it was called in Sumenep, a prahu with a single mast and a large deckhouse amidships; the *padewakang*, a large fully decked vessel with high topsides and two or three masts; and the *palari*, similar to the *padewakang* but smaller, with one or two masts and a hybrid rig.

These vessel types of Sulawesi origin were well established in the eastern islands by the mid-nineteenth century, although they did not completely displace indigenous and Javanese vessels. The largest type, the *padewakang*, was more restricted in distribution and was operated mainly from the small islands of Sapeken, in the Kangean group, and Guwaguwa, in the Sapudi group (see Map 4). Sapeken was colonised from South Sulawesi and it is probable that this was also the case with Guwaguwa.[4] Vessels from these two islands, as well as some from Ra'as, ranged very widely: from Wetar, to the northeast of Timor, to Surabaya, Semarang and Singapore. They carried a wide variety of goods but specialised in marine produce. Yet while the mariners of Sapeken, Guwaguwa and Ra'as contributed much to the reputation of the Madurese as seafarers, they were exceptional. Most contemporary Madurese vessels had a much more limited range and were conspicuous mainly in the waters around Madura, especially the Madura Strait.

The popularity of the vessels of South Sulawesi types in Sumenep was due to their greater economy by comparison with those built in Java or, for that matter, Madura itself. The problem was the price of teak, which had risen steadily throughout the nineteenth century.

---

[4] Although the people of Guwaguwa regard themselves as Madurese, the grave of a founding ancestor on the island is that of a personage who bore the title of *Karaeng*, the title for a noble among the Makassarese. The name of the island suggests a place with many caves (*gua*, 'cave'), but this does not fit with the physical reality of the place. A more likely etymology is that the name was derived from Gowa, the former kingdom of the Makassar area. Duplication of a noun results in a diminutive, thus the name could have meant 'Little Gowa'.

Indeed, a 1907 government report stated that the main reason for the decreasing output in vessel production from Sumenep was the "reduced opportunities for the pilfering of [teak]wood" as a result of more vigilant forest management in the Kangean reserves, with prospective owners consequently preferring to buy their vessels directly from Makassar, where the 'junglewood' used was of no interest to the government and could be had for no cost other than the labour of extracting it (OMW, 1907: 247). Again, it is instructive to look at the situation in Sapudi. In 1890, in the Sapudi archipelago (a group including Ra'as and Guwaguwa) there was a total of 237 trading vessels, of which 36 were *peduang* and the remainder a melange of Madurese, Javanese and Makassarese types. However, there was no shipbuilding, new vessels being purchased from other places (Verwijk, 1893: 243–4).

The picture that emerges from this overview of pre-twentieth-century Madurese vessels is thus a pastiche, from which it is hard to discern any distinctive Madurese tradition — if indeed any such tradition could fairly be said to have existed at all. This situation began to change in the late nineteenth and early twentieth centuries, when the Madurese developed the three new types of vessel with which they would become famously associated: the *golekan*, the *janggolan* and the *lete-lete*. Because these three latter prahu types were developed as late as the early twentieth century and were quite different from the types previously used by the Madurese, it may seem inappropriate to refer to them as 'traditional'. Nevertheless, I describe them so deliberately, taking into account the completely indigenous nature of the designs with no European influence in hull or sailing rig, their cultural significance for the communities using them, and the absolute adherence to these vessel designs until the abrupt changes which occurred with motorisation.

Each of these three types had its own exclusive range of distribution. For the *golekan*, the home coast was the western half of the north coast of Madura, for the *janggolan* the western half of the south coast, and for the *lete-lete* Sumenep and especially the eastern islands. Not only were the vessel types for these three coastal zones different, but there was also significant differentiation in the economic niches which they occupied. In presenting these three dominant twentieth-century Madurese prahu types, the approach is therefore through the three coastal zones as mentioned above, introducing not only the vessels but also the main places from where they were operated and outlining the economic niches that sustained these local fleets prior to their involvement in the Java Sea timber trade.

The Northwest Coast: *Golekan*

By comparison with the more densely populated and developed south coast, the north coast of Madura has a distinctly rural flavour. The few towns along the almost dead-straight coastal road are small and widely separated. The population is denser than it appears, however, with numerous agricultural hamlets in the hinterland. In the early twentieth century, this was still a remote area with very limited land transport services. Consequently, prahus remained essential for communication with the outside world. With the sea as their highway, the orientation of the people of the north coast has been toward the Java Sea and Kalimantan rather than toward the Madura Strait and East Java as on the south coast.

Nevertheless, the north coast is inconvenient for large vessels, with no natural harbours for protection against the rough seas of the northwest monsoon. Before the twentieth century, the most important port on the north coast had been Arosbaya at the northwest corner of the island. This had for centuries been a famous port, but trading prahus had ceased to call there after about 1900 because of the silting up of the Arosbaya river and competition from land transport coupled with steam shipping from Kamal to Surabaya (OMW, 1906: 15). Nowadays only small fishing craft are based at Arosbaya. In a few other places along the north coast there are small inlets — hardly more than creeks — that can afford shelter for several prahus. Elsewhere, vessels had to be hauled up on the beach during the northwest monsoon.

The prahu type that was developed along the western half of the north coast, the *golekan*, was a purely Madurese creation. In keeping with the need to be hauled ashore, the hull was lean and shallow-bodied. Although for its length its capacity was modest, the *golekan* was more than just a working vessel: it was also a cultural artefact and a work of art. The sharply raked and pointed stem and stern timbers (*lenggi*), the boldly sweeping sheer line terminating at each end in superbly fashioned scroll-shaped finials (*sompengan*), and the intricate ornamental and symbolic carving work were all standard features of the design. The hulls were always painted white, with a polychrome sheer stripe, and the upper portions of the ends, together with the

Photo 8 A traditional *golekan*, viewed from astern, at Telaga Biru, 1983. These beautiful prahus were then being marginalised by non-traditional and utilitarian motor vessels such as those visible in the background.

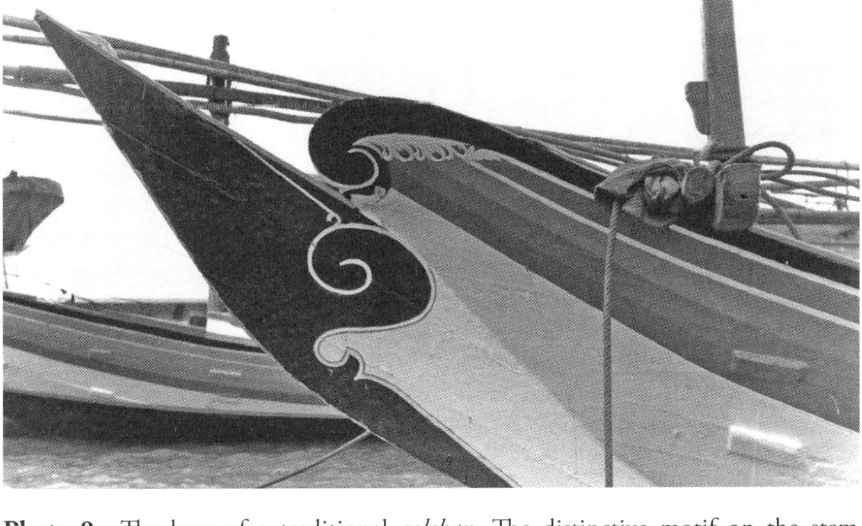

Photo 9 The bow of a traditional *golekan*. The distinctive motif on the stem and also at the stern is a standard feature of many Madurese prahu types. The protuberances on the planking (see also Photo 8) are the ends of thwarts which, arranged in tiers, reinforce the hull.

Photo 10 Boom crutch detail on a *golekan*.

Photo 11 Carving at the entrance to the deckhouse.

finials, highlighted in black. The general impression upon the onlooker was one of austerity and grace (Photos 8–11).

The aesthetic qualities of the *golekan* were not lost on C.A. Gibson-Hill, who described the prahu types he observed in Singapore shortly after the end of World War II. For Gibson-Hill, no other prahus could compare in beauty with these vessels of northwest Madura. In his description of the *golekan*, his often laconic style verged on the lyrical. Even the lateral rudder, devoid of embellishment but perfect in its studied asymmetry, moved him to write: "It is a lovely thing..." (Gibson-Hill, 1950: 126). Nor was his admiration of these vessels limited to their appearance: he was also enthusiastic about their behaviour in a seaway and their construction, both of which he regarded as superior to those of the Bugis *palari* (Gibson-Hill, 1950: 123, 126).[5]

Local fleets of *golekan*, used for both fishing and transport, were scattered along the northwest coast from Klampis to at least as far as Tamberu.[6] However, the two most important places were Sepulu and Telaga Biru, both in the kabupaten of Bangkalan. These two centres had the largest concentrations of vessels on the north coast, but what makes them of special interest for this study is their connection with the Kalimantan timber trade.

Sepulu

The small town of Sepulu, in the administrative sub-district (*kecamatan*) of the same name, had a population in 2002 of around 10,000 people. Although a single population cluster, it consists of two desa (in this context, administrative wards), Sepulu and Prancak. Such concentration of population is atypical of rural Madura but not unusual for coastal communities.

Being situated on an open coast with no inlet for shelter, Sepulu had few natural advantages for the development of a transport prahu

[5] In contrast to his high opinion of the *golekan*, Gibson-Hill regarded the *palari* as a rather ungainly craft. Half a century earlier the Dutch observer van Deventer had commented on the "cumbersome" build and performance of the predecessor of the *palari*, the *padewakang* (van Deventer, 1904: 110; see also OMW, 1907: 131).

[6] Until recently at least, the same name was also used for vessels found further to the east on the north coast, in Pasean and Ambunten. Although superficially similar in appearance, these craft differed considerably from the *golekan* of West Madura (see Appendix 1).

Photo 12 Small fishing outrigger canoe at Sepulu, 1983. The bow is to the right of the picture. The vessel in the background is a fish-carrying *golekan*.

fleet. Because of the need to haul the vessels up on the beach during the northwest monsoon, most boats were necessarily small to moderate in size. Formerly there were many small outrigger canoes there as well, used for trolling for bonito or similar species. They were usually referred to as *jukung polangan*, referring to the method of fishing (*jukung*, 'canoe'), but they were also called by the former generic name for outrigger craft, *peduang*. Some of these had exquisitely carved ends and in their traditional nautical artistry were without equal in the Indonesian archipelago. Since the 1980s a stone groyne has been built at each end of the beach to make a semi-sheltered harbour, but this is still inadequate against the northwest monsoon. During this season the larger vessels now based at Sepulu move to Kamal on the south coast, where they lie at anchor in the lee of the land on a tidal bay to the east of the ferry port.

In the late nineteenth century there were 270 trading vessels based at Sepulu, sailing mainly to Surabaya, Gresik, the island of Bawean in the Java Sea, and Kalimantan. By 1903 that fleet had already declined, presumably because of competition from the steam ferry at Kamal (OMW, 1906: 20). Nevertheless, as late as the early 1950s some *golekan* from Sepulu were sailing to Singapore with mixed cargoes of palm sugar, coffee, tamarind, dried fish, and mats, along with considerable numbers of passengers from Bawean (Piollet, 1995: 1972). This traffic to Singapore died out after 1953, although *golekan* still carried passengers and mixed goods between Sepulu and Bawean.

However, the mainstay of the economy of Sepulu was the fish trade. Besides many small fishing craft, there were also larger vessels specialising in fish transport. These fish transporters, medium-sized *golekan* about 12 metres in length with a large deckhouse, would sail far out into the Java Sea, even close to the coast of Kalimantan, and buy fish directly at sea from Bugis, Mandar and Madurese vessels. The fish would be salted at sea and, after being landed at Sepulu, would be dried for one or two days before being sent to Surabaya by truck. A few individuals became wealthy from this trade but since the mid-1980s the buying of fish at sea has been controlled by Banjarese entrepreneurs from Kalimantan, marginalising the more distant Sepulu merchants.

With the decline in long-distance fish trading, Sepulu's economy seemed to be facing stagnation. On the other side of the Java Sea, however, new opportunities were beckoning, especially in the Central Kalimantan town of Palangkaraya where there was a strong demand for labour. In keeping with its new role as provincial capital, this once small town was undergoing rapid development. Although situated far inland, Palangkaraya is accessible via the Kahayan river. In the mid-1970s *golekan* from Sepulu, fitted with motors to enable them to progress against the current, began to carry passengers from West Madura looking for work in this distant boom town. Although there had long been a trickle of passengers to Central Kalimantan, because of the new opportunities in the construction industry, this passenger service became a significant line of business. It was not long before new and larger *golekan*, fitted with engines and larger deckhouses to accommodate passengers, came into service to take advantage of this niche. Yet, despite this outward passenger traffic to Kalimantan, Sepulu remained until the late 1980s a sleepy little town tucked away from the main highway, dependent mainly on fishing and petty commerce. During the 1990s, however, as local entrepreneurs became involved in the timber trade, Sepulu would undergo a transformation to become the most dynamic place in Madura.

Telaga Biru

The next town to the east beyond Sepulu, Telaga Biru is part of the administrative sub-district of Tanjung Bumi. With around 3,500 households, the town is about the same size as Sepulu but there is only one desa (Telaga Biru). The modest size of the place belies its economic importance as it is the main market town for the hinterland and on market days the main road is extremely congested. It has also been, throughout most of the twentieth century at least, by far the most important transport prahu centre along the north coast. The *golekan* of Telaga Biru were both larger and more numerous than elsewhere and remarkably consistent in size — about 55 feet in length by about 14 feet beam. Like Sepulu, Telaga Biru is situated on an open coast with no inlet. However, a stone breakwater was built by the government to form an artificial harbour. According to a former desa head, the breakwater was built in 1936. Until the more recent (and still not fully adequate) work carried out at Sepulu, Telaga Biru was the only place along the whole north coast with a proper harbour, and until the 1990s, it was the only place on the coast with a port administration (*kesyahbandaran*) office. The building of the breakwater at Telaga Biru was not because of any natural advantages of the place and presumably was carried out because Telaga Biru already had a large enough fleet of vessels to warrant such expense.

Despite the prominence of Telaga Biru as a centre for transport prahus during the early twentieth century, and possibly a good deal earlier than that as well, the place is little mentioned in standard sources. Kuntowijoyo does not include it in his list of prahu centres for the north coast and mentions it only among his list of towns on the *south* coast from which cattle were exported to Java (Kuntowijoyo, 1980: 297, 357). Even the Inquiry into Diminishing Indigenous Welfare, in its 1906 report on transport in the residency of Madura, made no mention of Telaga Biru among the seaports of Bangkalan. The only ports of any significance noted for the *afdeeling* of Bangkalan were Arosbaya and Sepulu on the north coast, and Kwanyar and Blega on the south coast (OMW, 1906: 20). However, the records for indigenous shipping along the north coast of Bangkalan were destroyed in the early twentieth century by a fire in the district archive office and the apparent anomalies concerning Telaga Biru may be a consequence of that unfortunate event (OMW, 1906: 15).

In any case, the maritime entrepreneurs of Telaga Biru were apparently not much affected by the competition from land transport

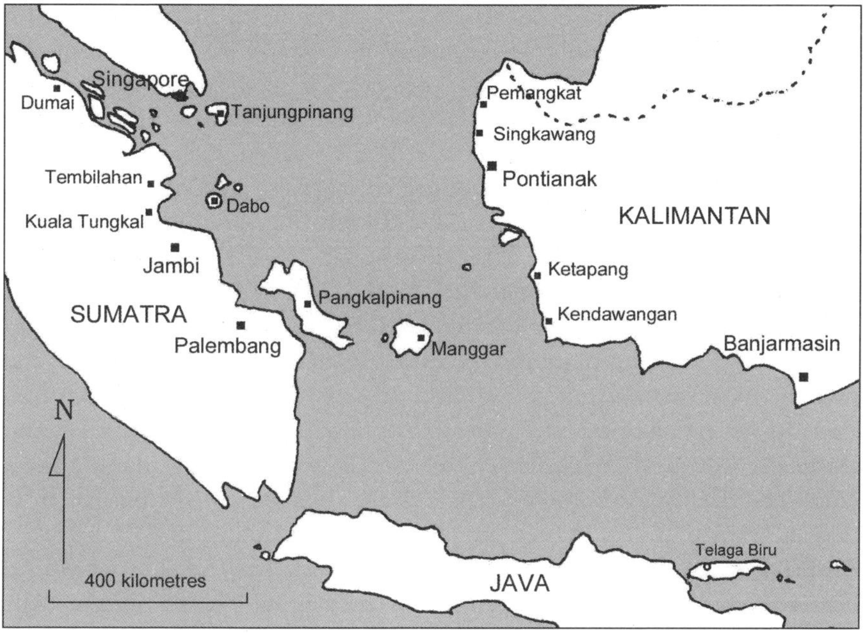

Map 7 Main destinations for *golekan* from northwest Madura

because their orientation was across the Java Sea rather than to the much nearer shore of Java. These were intrepid mariners, sailing wherever they might earn a profit, from Sumbawa in the east to Riau in the west, and also along the southern rim of Borneo and Sulawesi (Gibson-Hill, 1950: 121). Their specialty, however, was the western reaches of the archipelago: West Kalimantan, South Sumatra and Riau (Map 7). This strong association with the western part of the archipelago was a carryover from the traditional monsoonal trading pattern, with the prahus sailing swiftly and easily to the west with the southeast wind. These *golekan* carried a wide variety of goods, mainly produce from Madura: salt, cattle, goats, tamarind, palm sugar, dried fish and batik made by the women of Telaga Biru.[7] Another

[7] Telaga Biru is the only place in Madura where batik is produced, although formerly it was also made at Kedungdung on the south coast of Sampang. It is made only by the traditional method using the 'pen' (*canting*), not the 'mass production' stamp (*cap*) method now common in Central Java.

very significant 'cargo' was migrant passengers. Indeed, of all the prahu types of the archipelago, none has been more involved in the carrying of human freight than the *golekan*. As will become clear, the term 'human freight' was hardly an exaggeration to describe the nature of this traffic during the early decades of the twentieth century.

In the course of their trading voyages to distant places, the *golekan* skippers, always with an eye for profitable opportunities, became aware of the demand for labour in West Kalimantan. Local entrepreneurs, mostly Malays and Bugis who had already settled in these parts, were keen to recruit workers to clear land to develop rubber and coconut plantations. Hearing from these skippers of the poverty in West Madura, they offered to pay the skippers a commission for every worker they brought. Seizing the opportunity, the skippers recruited 'middlemen' to scour the countryside of West Madura. These middlemen regaled those they met with tales of how easy it was to make money in places like Ketapang and Pontianak. People who could not raise the fare were told that the skipper would bear the cost in the first instance and they would work off the debt soon enough.

The reality was much less pleasant than the migrants, mostly young single men and women, had been led to believe. Most of them had never been to sea before and the voyage must have been gruelling for them, especially in the cramped conditions with a hold filled with cargo, including livestock. On arrival they remained on board until prospective employers arrived, to whom they were 'sold' as contract labourers. To pay off their debt for the passage they had to work for up to two years clearing forest. This was a quite inordinate amount of labour to pay for the 'fare', with the benefits all going to the employer and the prahu skipper. Indeed, the carrying of passengers became such a lucrative business that some prahus sailed with passengers only, although most continued to carry goods as well (Sudagung, 2001: 77–9).

Madurese settlement in West Kalimantan started at Ketapang in 1902, while the first Madurese settlers to Pontianak came in 1910 (Sudagung, 2001: 78). Life was extremely hard for the migrants and many would have returned to Madura had it not been for the shame of returning in failure. Spurred on by their sense of pride, they redeemed their debts and relocated to new employers or to towns. Their keen awareness of their position as outsiders in the new land brought them together as an ethnic community, unlike in Madura where communality was almost non-existent. By helping and doing business with one another, coupled with hard work and frugality, many ultimately did well. In the 1930s there was a demand for workers on the main roads and many Madurese migrants took up this line of work. As stories of success filtered back, the flow of migrants from the impoverished hinterland of Bangkalan and Sampang increased, with new Madurese settlements being established further and further inland where there was more land available. This was especially so in the Sambas area, reached via the port of Pemangkat at the entrance to the Sambas river.

Spontaneous migration to West Kalimantan ceased abruptly with the Japanese invasion, although many Madurese were taken to Kalimantan as forced labourers *(romusha)*. In the 1950s spontaneous migration began to gather pace again, the main flow now being to the towns where there was much opportunity for unskilled people as road workers or as operators of trishaws (*becak*) or small ferry sampans. These migrants still came by *golekan* from Telaga Biru or nearby Banyuates. The economic mainstay for these voyages was cattle, with many skippers also the traders of the cattle. Sometimes with financial backing from others but often on their own behalf, skippers would buy cattle in the local market and sell them for a handsome profit in centres such as Pontianak, Manggar (on the island of Beliton) or Pangkalpinang (on Bangka). Some of the cattle would be sold for slaughter but others for animal husbandry. In either case, the buyers were typically migrant Madurese, who rapidly came to dominate the meat and cattle trades wherever Madurese settlements were established. The largest *golekan*, about 55 feet in length with a beam of about 14 feet, could carry up to 40 animals, tethered on a platform inside the long deckhouse. As well as the cattle, large amounts of fodder and water needed to be carried to nourish them for the voyage. Tending the cattle during the voyage was a considerable chore and often passengers who lacked money for the voyage would work their passage by carrying out this duty. The vessels would make about six voyages a year, during the dry season only, returning each time to their home port. Cattle were never carried during the wet season. With their highly strung nature, Madurese cattle are poor sailers and it was not rare for animals to die from shock in the event of an unseasonal storm.

From the 1950s until the late 1970s, there were between 150 and 200 of these large *golekan* based at Telaga Biru, with ownership widely distributed. A few owners possessed as many as four vessels but the majority of owners had only one. There was plenty of opportunity for work on the prahus during the sailing season, with each vessel having a crew of eight or nine. The *golekan* remained sail-powered until the mid-1970s, when the first engine was installed. At that time very few prahus had engines. Such motorisation as was occurring was mainly among Bugis vessels and this was taking place because consignors in main ports preferred to send their goods on a vessel with an engine, even though the freight rates were higher (Hughes, 1984: 187). In the case of Telaga Biru, the catalyst for motorisation was quite different and strictly local: the relentless outpouring of people from the hinterland of Madura in search of a better life.

For pure freight work, engines did not offer much potential for greater profit because for each round trip voyage, more time was spent in port than at sea. Any time saved at sea would not be enough to compensate for the high cost of the engine. It was better to put the cost of an engine toward another sailing vessel, which also offered insurance in the case of loss. Carrying of passengers, however, was a different matter. Prospective passengers naturally preferred to take a vessel going directly to their destination, so that those owners who could afford engines rapidly cornered the booming passenger market. By the early 1980s all the traditional prahus which had not been fitted with engines were hauled up on the beach, a forlorn spectacle. Most had bamboo mat covers to protect them against the sun, in the hope that they would eventually be fitted with motors and go to sea again. The last traditional *golekan* was built in 1983, yet even then the owner was waiting to raise funds in order to fit an engine before putting it into service.

Meanwhile, the prahus still in service had all been greatly modified, their owners having realised that adherence to tradition had no connection with profit. All these vessels had been given high and ugly superstructures extending for almost the whole length of the hull, while the decorative paintwork had been replaced with a drab monotone in any colour that happened to be available. There was not even a token sailing rig. The high superstructures were built to accommodate passengers, while livestock and salt could be carried below.

In fact, this carriage of passengers was in contravention of the shipping regulations, according to which long-distance passenger services should be provided only by approved modern ships. In the case of prospective passengers from Madura, this of course meant that they would need to travel to Surabaya first, entailing considerable extra

Photo 13 Some of the many engineless *golekan* hauled out on the beach at Telaga Biru in 1984, with matting covers to protect their decks from the sun. Most of these vessels never returned to service and were broken up after a few years. Only a single vessel survived in its original form and is now on display under a permanent shelter.

cost and trouble. Most therefore used the local service. In practice, traditional prahus often carried passengers, but only in small numbers. In the case of a larger group, the skipper would obtain a 'dispensation' from the local harbourmaster, with a payment as necessary to ensure approval. However, the scale of passenger transport from Telaga Biru went far beyond this modest 'rule-bending', with the larger vessels routinely carrying up to 150 passengers — without any proper life-saving or firefighting equipment on board, or even a radio. Such operations would not have been tolerated from the ports of Java but on the north coast of Madura the local authorities took no action.

In most cases nothing went wrong, however, and the transition to engine power was not marked by serious technical problems, unlike the case with South Sulawesi vessels.[8] In August 1983, however, a

[8] With the motorisation of the Bugis fleet, which took place from the mid-1970s until the early 1980s, there were widespread structural failures due to the thrust and vibration of the engines and many vessels sank (Amarell, 1999: 264, n.10). This structural problem was soon overcome by reinforcing the hulls with numerous steel bolts.

Telaga Biru vessel sank in rough weather in the Java Sea and 76 out of the 171 people on board were lost (Hughes, 1984: 217). Despite the scale of this tragedy, vessels continued to embark passengers from Telaga Biru and Banyuates in contravention of the shipping regulations. During the late 1980s some vessels from Telaga Biru also started carrying passengers to Riau, the gateway to Malaysia, where high earnings could be obtained in plantation and construction work. Such work was in most cases without official approval but, because of the demand for labour in Malaysia, the authorities there did not take a tough stance.

Although some operators in Telaga Biru were profiting from this passenger business, the 1980s was generally a time of declining opportunity and a widening of wealth differences. Whereas there had previously been a large fleet of sailing vessels, there were now fewer vessels, all with engines and of much larger size. As well as the inevitable fall in employment opportunities, the cultural vibrancy of the village seemed to have been sapped. Instead of the superb spectacle of the large fleet of *golekan*, of such uniform size and styling that they looked as if they had all been created by the same hand, Telaga Biru was in the mid-1980s an uninspiring place. The traditional prahu economy was stagnant, the few vessels in the port were grotesque with ugly superstructures imposed on the traditional hull form, while along the beach once proud *golekan* were being broken up for their teak timber or being used as latrines. The benefits of the *golekan* to the community of Telaga Biru had always been more than purely economic: the traditional vessel had given a sense of cultural pride and identification, almost like a totem, and this had been reflected in the care that owners devoted to their craft. That sense of communal pride had declined, a casualty of economic rationalism.

Although opportunities for work crewing on vessels had declined, a new line of work appeared, in the unloading of timber carried by the motor vessels on their return from Kalimantan. This timber was really only a sideline to the main business of carrying passengers and, to a lesser extent, cattle. Since the late 1960s, some *golekan* had been carrying timber on their return voyages, discharging this at Jepara or Semarang before continuing back to Madura. During the 1980s, however, increasing amounts of timber began to be brought back to Telaga Biru. It was intended mainly for local consumption, in the new houses and extensions that the more successful vessel owners could now afford. In the 1990s this modest local timber trade would boom, with trucks from centres throughout East Java and even Bali regularly making their way to Telaga Biru to obtain the high-quality timber being landed there.

The Southwest Coast: *Janggolan*

The most westerly part of the south coast of Madura, from Kamal to Kwanyar, is a very short distance from Java, so that the orientation along this stretch of coast is very much toward the city of Surabaya. Nevertheless, the main road from the port of Kamal does not follow the western half of the south coast but instead leads north to Bangkalan before turning southeast toward the town of Sampang. Consequently the western half of the south coast is more isolated than might be expected from its proximity to Java, while the more distant part of this coast, leading up to the western shore of the Blega estuary — the area of interest for this study — is very isolated.

The traditional prahu of this southern coastline of West Madura, the *janggolan*, was a true freight vessel and, as such, larger and more powerful than the *golekan*. Gibson-Hill (1950: 124) considered it "less lovely" than the lean and elegant north coast craft but no less singular. Indeed, the *janggolan* was perhaps the most remarkable, and certainly the most distinctive, of all modern prahu types. Structurally it was related to the *golekan* but the general appearance was very different. The hull form was unique among modern transport prahus, with the keel timbers projecting well beyond the hull proper at each end and the planking above terminating against narrow flat transoms. Because this resulted in a bifurcated profile at the ends, compared to the straight or smoothly curved profile of most vessels, the *janggolan* can be classed as a 'bifid-ended' type of vessel (c.f. Hornell, 1970). The rig was related to that of the *golekan* but more highly developed, a complex arrangement of sails, poles and lines incomprehensible in its operation to those without direct experience of it. Finally, despite its purpose as a dedicated cargo carrier, there was nothing utilitarian about the appearance of the *janggolan*, with profuse ornamentation an integral element of the design. Especially important in this respect were the bow and stern transoms, which were covered with carved motifs arranged in a particular pattern, highlighted with polychrome paint. Like the *golekan*, the *janggolan* was a cultural icon, the most outstanding example of the material culture of the coastal people who relied upon it and of deep symbolic as well as economic importance to them. No part of it could be changed without affecting the whole.

Photo 14 A large *janggolan* laid up for the northwest monsoon at Batuputih, Sreseh, in January 1998. The yards and booms have been temporarily removed. Launched in 1995, this was perhaps the last large *janggolan* to be built. The gantry-hung rudders, nearly seven metres long, are a departure from the standard arrangement in which the rudders were stowed on deck when not in use (c.f. Photo 25).

Photo 15 *Janggolan* under sail, en route from Beliton to Java, 1984. The stout beam extended well outboard is to assist in staying the rig. A similar beam is extended on the other side. The decorative paintwork and the carvings on the transoms at each end of the vessel are standard features of the design.

In the early twentieth century, *janggolan* were a common sight in Surabaya. Fleets of these striking vessels were operated from Kwanyar and Sukolilo Barat on the southwest coast of Madura and were especially used for ship-to-shore work in the busy Surabaya roads. However, because of the increasing competition from motorised harbour vessels, coupled with the Kamal ferry service and better land transport services on Madura, the transport fleets of southwest Madura dwindled away, disappearing altogether during the 1970s. Sukolilo Barat and Kwanyar are now only fishing ports but the fishing fleets in both places are comprised entirely of small bifid-ended vessels of the same generic type as the *janggolan*, called *alis-alis* (see Appendix 1).

Further east, in the area bounded by the Sampang and Blega rivers in the kabupaten of Sampang, the *janggolan* survived virtually until the end of the twentieth century, with many built even into the 1990s. One fleet of these vessels was based in the Sampang river, which provides access for shallow-draught vessels from the south coast to the town of Sampang. Sampang was of old a centre for maritime trade, with a good deal of prahu traffic between there and the ports of East Java, especially Pasururan and Probolinggo. Like Kwanyar, by the late nineteenth century it was already in decline as a maritime trade centre and the opening of the railway line, followed by bus and truck services, accelerated this trend. The Sampang fleet of *janggolan* disappeared in the late 1980s, while another smaller fleet based at the nearby island of Mandangil (see Map 3) also dwindled away around this time.

Apart from the Sampang river, the other main place for *janggolan* was the broad estuary of the Blega river, a short distance to the west (Map 8). This river, the largest waterway in Madura, is referred to by locals as the *sungai rejo* ('great river') — although it is hardly a great river by standards in Java, let alone Kalimantan. From the estuary the river leads inland to the small market town of Blega, once a trading port but long not visited by prahus. Nevertheless, a substantial fleet of transport prahus has continued to be operated from the desa of Sreseh on the western shore of the estuary, as well as from the adjoining desa of Taman on the south coast. Both desa are in the kecamatan of Sreseh, with its administrative centre at Labuhan.

Until the mid-1990s the Sreseh fleet consisted entirely of *janggolan*, none of them with engines. By this time most prahus elsewhere were relying on engine power but the deeply traditional character of

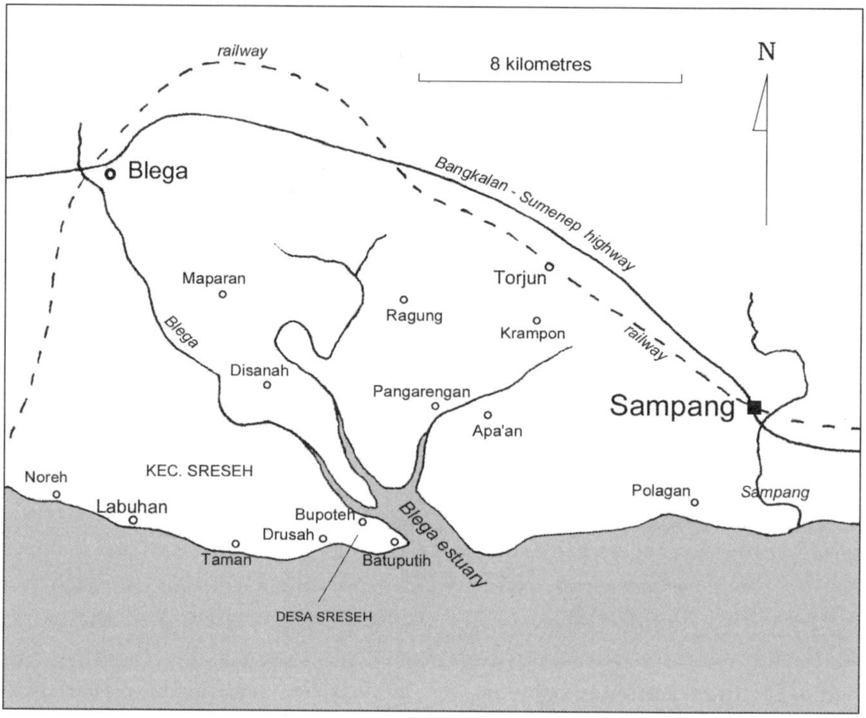

Map 8 The Blega estuary and surrounds

the *janggolan* was inconsistent with the rational business approach that has usually accompanied the installation of engines in prahu fleets. Some skippers felt that the elaborate ornamentation, involving much labour and expense, was no longer appropriate in the modern world, while the stern profile was also inconvenient for the installation of a propeller. Others countered that changes to the traditional design that had served so well would result in bad luck. In 1995 a Sreseh man built a vessel of utilitarian design, based on the type of prahu used by Mandar vessel operators, and installed an engine. Thereafter all new vessels built were of this type, called *sepel*,[9] while the *janggolan*, the vessel with which this community had so strongly identified, so rapidly became regarded as obsolete that by 2002 none were left in service.

[9] Pronounced 'se-pell', with the emphasis on the second syllable.

Photo 16 A *sepel*, the type that became popular in Sreseh after 1995, anchored off Drusah in early 1998. The advantages of the *sepel* over its predecessor were its lower cost, easier construction of the ends, the shelter of the after cabin for the crew and quicker tacking. Sailors were nevertheless in agreement that the *janggolan* was the faster sailer.

The desa of Sreseh is an unexpectedly remote place. Until the mid-1980s there was virtually no road transport beyond Labuhan, and until 2000 the road to Batuputih at the end of the peninsula was a rough track, often difficult to negotiate in the wet season. There is no local market, the nearest ones being at Labuhan to the west and Pangarengan on the eastern shore of the estuary. The sense of isolation is enhanced by the absence of any large clustered settlement. The population of around 1,100 households is highly dispersed across hamlets in the pattern typical of rural Madura. The largest hamlets, Batuputih and Bupoteh, are strand settlements while the desa head resides in a tiny inland hamlet. For most of the population, access to the outside world is by a *sampan* ferry service to Pangarengan, at the head of a small creek on the eastern shore of the estuary from where motor vehicle transport is available to the town of Sampang. In earlier times prahus were frequently used for travelling to Java and this is reflected in the strong social ties between Sreseh and the towns of Pasuruan and Probolingo on the far side of the Madura Strait.

By any standards, Sreseh qualifies as a backward village (*desa tertinggal*). Most households are fairly poor and in 1999–2000 there was still not a single person in the desa who had undertaken tertiary education. There were a few fishing boats in and near the estuary, although there were larger and obviously more affluent fishing villages at Labuhan and Noreh, about ten kilometres to the west. Agriculture is subsistence only, mostly maize, and even then not nearly enough to meet local needs. The local economy has instead been overwhelmingly dependent on the production and movement of Madura's most important commodity: salt (see Chapter 2).

The state salt enterprise set up by the Dutch in Sampang was concentrated at Pangarengan and Ragung on the eastern shore of the Blega estuary, a natural choice because of the extensive swampland there. Much of the salt went to the rail siding at Torjun, a considerable distance from the lower estuary, but sea transport was also important. For the vessel operators of the neighbouring strand hamlets, there were two lines of work associated with salt. The first was lighterage, carrying salt from the packing houses at Pangarengan and further upriver to ships anchored offshore. The Blega estuary is a very shallow place, with a broad sandbank extending well offshore from the mouth of the estuary and only a narrow tidal stream running to the salt storage sheds. Ships based at Kalianget, where the state salt enterprise had its headquarters, used to call regularly to Sampang to load salt but because of their draught they had to anchor far out, near the island of Mandangil. This work was often hampered by rough conditions, while in the mornings of the dry season the boisterous southeast wind was a direct headwind for prahus coming out of the estuary. On the other hand, being appreciably closer to Surabaya than the other salt-producing areas of Madura, the Blega estuary was favourably situated for the second line of prahu work — direct transport of salt to the ports of East and Central Java (van der Kemp, 1894: 267; OMW, 1907: 235). The most common destinations were Pasuruan and Surabaya but the larger prahus routinely carried cargoes of salt to Rembang and Jepara in Central Java.

On these longer passages, the carrying of salt would often be combined with other coasting work. In 2000 the oldest former prahu sailor in the desa of Sreseh was an 83-year-old man named Maderai from the hamlet of Drusah. Maderai's sailing career had started during the colonial period. He related that the *parao janggol* (*janggolan*) he worked on would sail with a cargo of salt from Sampang in May, after the southeast wind had set in, bound for Semarang. After the salt had been discharged, they would take on a load of sugar cane and sail to Lampung in southern Sumatra, where the cane would be unloaded. From Lampung they would load charcoal, which they would carry to Batavia, before returning empty to Madura to take a second load of salt. Rather than Sampang, however, this second load was from Kalianget. They would then sail again to Central Java, going via the north coast of Madura. After discharging the salt they would return to Sreseh and lay up the prahu for the wet season.

After the disruption of the war and the departure of the Dutch, the local prahu operators benefited from the decline in modern shipping services, so that *janggolan* continued to be heavily involved in the transport of salt. This role became even more important after deregulation of the industry in 1959, with local entrepreneurs developing extensive salt holdings in the swamplands of the river delta. This area is bounded entirely by water and even today there is no road access to the villages of Soro'an, Disanah and Maparan, which were built there after the conversion of the land for salt production. The prahu operators of Sreseh were ideally placed to take advantage of these new saltlands and they could now deal directly with the salt producers. A small number of private salt producers with large holdings became wealthy, more so than any of the prahu owners. Some smaller salt producers had prahus built for their own use, so that they could transport their salt to the market at Pasuruan without paying for freight or waiting for a vessel to become available, while some of the more successful prahu operators diversified into salt production. Some enterprising vessel owners who had a reputation for integrity even worked as agents, simplifying the problems of transport and sale for the small private producers. Larger vessels sailed regularly with salt to the prahu harbour of Kali Baru, near Tanjung Priok, Jakarta, and continued to do other cargo work during the sailing season as before the war, while for the smaller prahus Pasuruan became virtually the sole destination and salt the only cargo.

Despite the efficiency of the salt transport service provided by these sailing vessels, they started to become marginalised as the economy surged ahead in the late 1970s. Truck transport became increasingly competitive with prahus and this competition became overwhelming with the establishment of new salt-receiving and processing centres at Surabaya, Pamekasan and Camplong, not far from Sampang. All salt from the state salt corporation's holdings in Pamekasan and

Photo 17 A small salt carrier returning to the Blega estuary after a delivery to Pasuruan. This vessel would have carried about 35 tons of salt, with little reserve freeboard. The hold is undecked for convenience in loading and unloading the salt, using manual labour.

Sampang began to go to these new receiving depots, eliminating the lighterage work that had been a valuable standby for local prahu operators. Worse, the private salt producers from Polagan, Apa'an and Pangarengan on the eastern shore of the Blega estuary also began to transport their salt by truck. This was convenient for these producers, because there was good road access through this area. For the private salt producers of the western shore, in the desa of Sreseh, however, truck transport involved transhipment to a depot with land transport access, which significantly raised the cost. Moreover there was strong mistrust of the salt cooperative, through which salt could be channelled into the state system, with accusations of routine cheating by those in charge. For all these reasons, the private salt producers of the western shore have continued until the present to send their salt to Pasuruan, using local prahus. As late as 1998 most of the vessels in this salt trade were relying on sail alone but now all have engines.

Janggolan continued to carry salt from Sreseh to Jakarta until the early 1980s, when this long-distance niche could no longer be maintained in the face of competition from road transport and salt produced at centres on Java, especially Rembang. The decline in the long-distance salt trade was not a serious matter for the maritime entrepreneurs of Sreseh, however, for since the late 1960s many of them had been involved in the timber trade. Even as the salt trade began to wane in the late 1970s, it was apparent to owners of large *janggolan* that fortunes could be made in the Java Sea timber trade.

Sumenep and the Eastern Islands: *Lete-lete*

Although Sumenep, at the eastern end of Madura, is much further than Bangkalan and Sampang from Java, any notion that it might therefore be 'more Madurese' than the centres of West Madura would be wrong. In fact, external influence has been much greater in Sumenep than anywhere else on Madura because Sumenep was favourably placed for trade not only with the eastern end of Java but also with the Lesser Sundas, Sulawesi and Kalimantan. Because the town of Sumenep is situated inland with no navigable access, the main port has always been nearby Kalianget. Whereas the other centres of the south coast of Madura were little involved in trade outside the Madura Strait, Kalianget was the most important port on Madura for long-distance trade and also a transit port connecting Java with the eastern islands of Madura (Kuntowijoyo, 1980: 128).

What was especially remarkable about Sumenep was its high proportion of 'foreign easterners', as they were categorised: mainly Chinese, Arabs, Malays and Bugis. In 1815 nearly 6 per cent of the population of Madura were such 'foreigners', two-thirds of whom resided in Sumenep. Apart from Batavia, nowhere else in all of Java and Madura had such a high proportion of outsiders (Raffles, 1978 [1817]: vol. I: 63; vol. II: 284–6). The Chinese were especially prominent. An enduring testament to their influence can be seen on the front wall of the Sumenep main mosque, which is extensively decorated with a key-like Chinese motif called *kihong*, a term possibly derived from *Tjionghoa* ('Chinese').

The Chinese and Arabs lived mostly in the town or port area, while the Malays and Bugis tended to settle either on the coast or on the offshore islands, especially in the Kangean group. The 'Bugis' also included Mandar and Makassarese settlers; all three groups appear to have been listed at times by the Dutch under the ethnic label of 'Malay' (Kuntowijoyo, 1980: 81). These 'foreign easterners' were much involved in trade and, until the early nineteenth century, some of the Malays, Bugis and Madurese of the offshore islands were also involved in piracy and smuggling (Raffles, 1978 [1817], vol. I: 57). The Dutch presence in Sumenep was also unusually large because of the salt industry in Kalianget and traces of Dutch influence are discernible in the local maritime culture.[10] Although all these outsiders constituted a minority, their significance was disproportionate to their numbers, especially in the town and coastal areas. Consequently, Sumenep was the 'melting pot' of Madura. Over time, most of these outsider groups have merged into the mainstream of Madurese society but their cultural influence remains.

This diversity of influence was apparent in the prahu types of East Madura at the start of the twentieth century. On the south coast there was the *kacik*, based on the *pencalang* or *cemplon* of Java; on the north coast and also in the eastern islands was the *golekan*, similar in form to the vessel of the same name in West Madura but smaller and of different construction, and often as an open boat; while in the eastern islands, there were many craft of South Sulawesi origin, with some of the largest type in this category, the *padewakang*, having a capacity of up to 50 tons.

However, during the next two decades, the Sulawesi vessel types disappeared from the eastern islands, while the *kacik* also began to decline in popularity. A new hybrid type known as the *lete-lete* gained the ascendancy and remained the dominant transport prahu of Sumenep until almost the end of the twentieth century. The *lete-lete* became so popular that around 1930 it also spread to the west coast of South Sulawesi. Nooteboom (1940), who recorded the type on the Mandar coast, noted that it was not a local development and had been introduced from Madura. The *lete-lete* bore a superficial resemblance to the *golekan* of the northwest coast, but it was more burdensome

[10] Three examples come to mind. First, the very heavy lodging (horizontal) knees bracing the hulls of some transport prahus of Sumenep, which are not a feature of traditional prahus of anywhere else; second, the heavy square-sectioned rail timber (*pordu*) of the fishing vessels of Sumenep, again not a feature of other traditional Indonesian watercraft but corresponding to the *boord* in traditional Dutch craft (see Huitema, 1995: 37); third, the rudder of the European-styled *sekoci* (named after the Dutch vessel type, *schuitje*) of Ra'as was in the classic Dutch style and quite different from other centrally mounted rudders in Indonesia during the twentieth century.

Photo 18 A *lete-lete* at low tide at Prenduan, on the south coast of Sumenep, 1983. The ability to take the ground is essential for Madurese prahus. Despite their shallow draft, these vessels are good performers to windward, lateral resistance being provided by the lateral rudder, which functions much like a leeboard.

than the latter and also completely different in construction and sailing rig. It was a rugged and capacious craft, well suited to its purpose as a dedicated small cargo vessel, and its hull form, at least, would not have looked out of place in the eastern Mediterranean.

Although substantial fleets of transport prahus were also based on the mainland of East Madura, these have now all disappeared apart from a few vessels based at Pasian and Ambunten on the northeast coast. Since the early twentieth century, most of the transport prahus of East Madura have instead been based in the eastern islands. Prahus remained more important in these islands for the obvious reason that the islanders were more dependent on them for their transport needs and because of the lack of alternative livelihoods. Most of the eastern islands are too small to have a strong agricultural base and the limestone soil is poor. Living standards on these islands were especially harsh during the early decades of the twentieth century: in 1933 the

4,000 or so residents of Giliiyang all but starved for three months (Kuntowijoyo, 1980: 102).

All of the inhabited eastern islands had prahu fleets but the most important have been those of Giligenting, Sapudi, Guwaguwa and Ra'as. However, prahus from Ra'as are no longer involved in transport work, while most Guwaguwa prahus work the coasting trade around Kalimantan, with only a few involved in the Kalimantan–Java timber trade. This study will therefore focus only on the islands of Giligenting and Sapudi.

Giligenting

This small island, only about 15 square kilometres in area, lies a short distance off the Saronggi peninsula in the southeast of the mainland of Sumenep (Map 9). The most important prahu-building centre in all Madura is situated in this peninsula, in the village of Kebun Dadap. Although Giligenting prahus have in recent decades been built mainly on the island, in earlier times many were purchased from Kebun Dadap. Administratively, Giligenting is part of the kecamatan of the same name, which also includes the neighbouring island of Giliraja (see Map 4). Because *gili* means 'small island' in Madurese, the names of many islands around Madura bear this prefix. Giligenting is densely populated with over 900 persons per square kilometre — the highest density for the whole kabupaten of Sumenep with the exception of the urban areas of Sumenep town and Kalianget (BPS Sumenep, 2000). The island is divided into four desa: Aeng Anyar, Bringsang, Gedugan and Galis, with transport vessels based in all four. At Aeng Anyar there is a concrete jetty, with prahu ferries to Tanjung on the adjacent mainland. The coastline is mostly sandy with ample shelter from prevailing winds, so that places abound for the beaching of vessels for maintenance or for building.

In 2003 a remarkable feature of Giligenting was the many large and nearly new but unoccupied houses, reflecting the highly outward-oriented economy. These houses were in the main owned by local people formerly involved in the timber trade in West Java but now operating foodstalls in Jakarta. Despite having worked for years, even decades, in West Java, their prime attachment is still to their native island, to which they return home each year during the latter part of the fasting month. When they eventually retire, they plan to live in Giligenting.

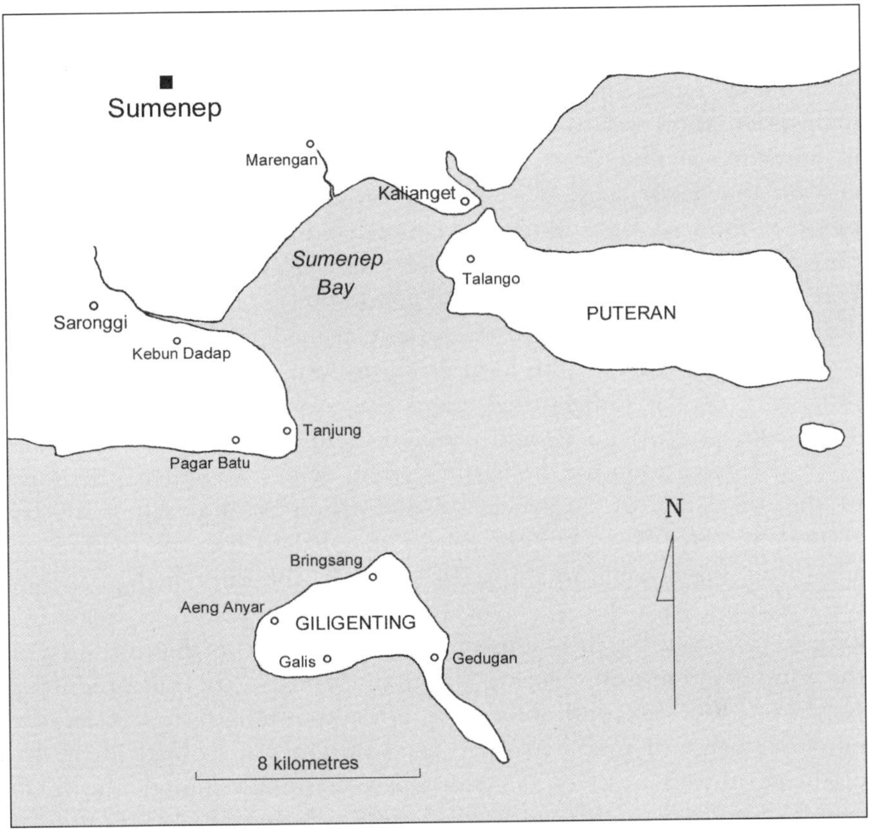

Map 9 Sumenep Bay and Giligenting

Since the early twentieth century, Giligenting has had the largest prahu fleet and the most successful maritime economy of all the eastern islands of Madura. Until the transformation which occurred with motorisation of this fleet in the 1990s, Giligenting *lete-lete* — easily recognised by a stoutly constructed 'doghouse' abaft the main gabled deckhouse — were a common and distinctive sight in ports all around the Java Sea, from Sumbawa to Riau. With motorisation and the building of new vessel types, vessels from Giligenting have been less conspicuous, blending in with those from other places, but they are still the most common of the Madurese transport vessels.

The factors underlying the success of the maritime entrepreneurs from this small island are less obvious. At the start of the twentieth century there were many Sumenep prahus sailing to Banjarmasin,

Palembang and Riau among other places, involved in trade on their owners' account (OMW, 1907: 130). Prahus from Giligenting were presumably represented among these long-distance sailers but their main orientation was to Java. The Giligenting fleet was not the largest in Sumenep at that time, larger fleets then being based at Kalianget and on the north coast.[11] This situation changed with the establishment of modern land transport, coupled with modernisation of the port of Kalianget and a motorised ferry service to the nearby island of Puteran. The vessel operators of Giligenting were less affected by these developments because prahus were still needed to service the island economy. Nevertheless, the local economy was small. The continuing prosperity of the Giligenting prahu operators clearly depended on more than carrying goods and people to and from their home island.

The great boon for the island's prahu owners was their proximity to the salt lands of Sumenep. Madura's best natural salt lands lay around Sumenep Bay, but the creeks that led to the salt production areas were inaccessible to ships. To load salt, the ships had to anchor well offshore, with the salt being brought out by prahus or lightering barges. However, because Sumenep Bay is exposed to the full force of the southeast monsoon during the dry season — the salt-producing season — the sea conditions were often too rough to transfer the salt from barge or prahu to ship. The solution was to load in the lee of the northwest coast of Giligenting, so naturally enough the small vessels employed in this work were from Giligenting. An account of this arrangement is provided by the Englishman G.W. Earl, who was on board a square-rigged sailing ship that loaded salt from Sumenep in November 1832. The loading was organised in advance, so that the morning after Earl's ship anchored in the lee of Giligenting "several large vessels" sailed out laden with salt for transshipment (Earl, 1971 [1837]: 75).

Apart from salt lighterage work, Giligenting prahus carried salt to ports on the other side of Madura Strait, where the population was growing rapidly as a result of expansion of the sugar industry. Shipping of salt over long distances was undertaken by European sailing

[11] Among trading prahus of Sumenep administrative district in 1903, 340 were based at Sumenep (Kalianget), 440 in the Northeast district (the north coast and Giliiyang), and 237 in the Southeast district, which included Giligenting (OMW, 1906: 20). These figures did not include Sapudi and Kangean, which were treated as separate districts.

ships but it was hardly economic to use these ships to transport salt across the Madura Strait, where the ports were too shallow for the ships to be able to enter, so that lighters would again have to be used. Consequently, this short-haul salt trade devolved upon local prahus. Much salt was carried by prahu from the narrow inlet of Marengan in Sumenep Bay, where the heavily laden prahus were just able to scrape out of the tributary creeks with the assistance of a good high tide. From at least the late nineteenth century, there was a brisk traffic of salt-carrying prahus from Marengan to Surabaya, Pasuruan, Probolinggo, Besuki and Bali (van der Kemp, 1894: 273). This was hardly lucrative work, with the Chinese salt transport contractors keeping their expenses low by making advance payments to vessel owners, thereby keeping them in their debt (Kuntowijoyo, 1980: 292). Probably the Giligenting prahu owners were more easily manipulated in this respect because of their lack of alternative work. According to the retiring desa head of Aeng Anyar in 2002, a man who also owned four prahus, salt was formerly "the only cargo" for Giligenting vessels. This steady salt transport work enabled the Giligenting entrepreneurs to cope during the early decades of the twentieth century, when conditions were generally difficult for indigenous shipping, and it enabled them to survive while other fleets dwindled away. They were thus well poised to take advantage of the greatly increased freight opportunities for prahus during the depression years of the 1930s.

Between 1947 and 1950, when the salt monopoly was disrupted, *lete-lete* from Giligenting carried salt from Jepara to Sumatra, Riau and even Singapore (Piollet, 1995: 80). The capacity of these long-distance *lete-lete* was around 50 cubic metres, a large increase on the average size at the start of the century, reflecting their employment in carrying rather than trading. This capacity was similar to that of contemporary Bugis vessels, although the Bugis vessels appeared the larger because of the tall masts and long bowsprit of their European gaff ketch rig.[12]

After Independence, Giligenting vessels were again able to take up long-distance salt transport work from Kalianget, stepping into the role formerly undertaken by ships. During the 1950s West Kalimantan was a common destination for these salt-carrying prahus. Apart from Kalianget, Giligenting vessels often loaded salt from Bunder, in

---

12 This statement is based on dimensions for both types in Gibson-Hill (1950: 112, 128).

Pamekasan, and from the Sampang river and the Blega estuary. Salt remained an important cargo until the 1970s, when small steel bulk carriers based at Kalianget took over the long-distance salt freight work. In 2003 salt was still being shipped by prahu from Kalianget to Panarukan in East Java but the vessels doing this work were mostly from Kalianget, not Giligenting. As at Sreseh, this was by then very much the poor end of the prahu transport market.

Important as salt was to the rise of the Giligenting fleet, it was far from the only important cargo line for these vessels prior to the timber boom. The *lete-lete* was an optimal design of small sailing cargo vessel for Indonesian waters, stoutly built and with great load-carrying ability, and it routinely carried a wide variety of bulk goods, from sand and coral stones to crates of bottled soy sauce. With its broad beam and large deckhouse, the type was also well suited to carrying livestock (Dick, 1975a: 74). Before the timber boom, two of the most important lines of work were the shipping of cattle from Sumenep to Banjarmasin, and horses from Lombok and Sumbawa to Java. Both of these trades faded out in the middle of the twentieth century. Another economic standby was the shipping of onions and garlic, again from Lombok and Sumbawa, to Kalimantan.

Sapudi

The largest of the inner islands of Sumenep, Sapudi has an area of approximately 130 square kilometres (Map 4). In 2000 the population was nearly 48,000, about three times that of Giligenting but much more dispersed (BPS Sumenep, 2000). There are two small towns, Gayam on the south coast and Nonggunong on the northwest coast, and the island is divided into two kecamatan of the same names. Gayam is the more important and also the port for the ferry from Kalianget, a journey that takes four hours. Many people prefer to take a prahu from Dungkek, at the northeastern tip of Madura, because the journey is shorter. However, the Dungkek prahu ferry is small with no comforts and, on arrival at Sapudi, the passengers have to wade ashore on a rough beach on the western shore before catching a 'pick-up' utility vehicle (*pikup*) to elsewhere on the island.

Despite its considerable distance from the 'mainland' of Madura, Sapudi has strong historical associations and is well known for its ancient relics. The apparent remoteness of the island is a modern perception that disregards the ease with which mariners were able to

Photo 19 A *lete-lete* from Giligenting, loaded with drums of kerosene, in a creek near Bunder, Pamekasan, in 1983. Going to sea in such a heavily laden condition was routine for these prahus. The deck is watertight and the deckhouse structure strongly planked, like the hull.

traverse the highway of the sea in pre-modern times. It is therefore no anomaly that Sapudi is regarded as a heartland of Sumenep culture. The legendary figure Adipoday, putative father of Madura's great folk hero, Jokotole, once ruled here and held sway over all the other islands thereabouts. So great was his influence that many locals even today refer to Sapudi as 'Poday'. His grave in the old village of Nyamplong in the west of the island is still regarded as a site of spiritual power, with people often travelling there to meditate and pray (*berziarah*). One of Adipoday's legacies is the stands of nyamplong (*calophyllum inophyllum*) trees for which the island is known. According to local tradition, he encouraged the planting of these extremely useful trees (Abdurachman, 1971: 12). Apart from its edible fruit which can also be processed for oil, the branches of this tree are excellent for crooks for the framing and reinforcing of prahus.[13] It is also an exceptionally

[13] Piollet (1995: 86) notes that the timber used for this purpose on Sapudi is *bintangor*, but this is another name for the same species (see Perla, 2002: 30).

good timber for dugout canoes because it is light yet resistant to splitting as a result of its interlocked grain.

Another reputed legacy of Adipoday is the cattle-rearing culture for which Sapudi is famous. Indeed, one of Madura's best-known cultural institutions, its bull races, are widely regarded as having originated here, and the winners of the all-Madura finals have more often than not been from Sapudi (Abdurachman, 1971: 12).With ample area in the hilly interior for the collection of fodder, cattle-rearing is the most important economic activity on the island. According to the current head of the desa of Gayam, many people in the interior have been so successful from cattle-rearing that they have been able to afford the pilgrimage to Mecca. Nevertheless, this business success is not apparent from the houses. The traditional Madurese values of thrift and disapproval of conspicuous consumption remain strong on Sapudi, so that the ostentatious houses so common in suburban Java are not as yet a feature of the local cultural landscape.

The cattle industry provided a natural economic opportunity for the vessel owners of Sapudi and the capacious and stable *lete-lete* was well suited to this line of work, with the orientation as much toward Kalimantan as to Java. Horses and goats were also carried, as well as a range of other agricultural products, coral stones and slaked lime for use in mortar and plastering. In the early twentieth century the prahu economy of the Sapudi islands, a group including Ra'as and Guwaguwa as well as Sapudi itself, was the most vigorous of anywhere in Madura. In 1895 there were 286 transport prahus in the Sapudi islands. By 1903 the number had risen to 440, an increase of 51 per cent in just 8 years. This large increase may have been due to more than buoyant trading conditions, however. The waters around the outer islands of the Sumenep archipelago were formerly a haunt of pirates and one of the reasons cited for the rapid increase in numbers of Sumenep vessels around this time was the "improved security at sea" (OMW, 1906: 14, 20).

The range of operations for Sapudi vessels has been just as wide as for those of Giligenting, except for a lesser involvement in the Madura Strait area. Sapudi *lete-lete* were formerly common visitors to Singapore and, according to Piollet (1995: 90), even to the port of Medan (Belawan) in North Sumatra. In recent decades the carrying of cattle from Sapudi has declined in importance largely because the cattle-rearing industry has become well established in South Kalimantan, where the industry is dominated by migrants from Sapudi

Photo 20 Two powerful *lete-lete* from Sapudi, beached for maintenance at Sepulu, 1983. In the right foreground is the stern of a small *golekan*.

and other parts of Madura. Many prahus from Sapudi and nearby Guwaguwa then resorted to carrying sand around the coast of South and East Kalimantan. Frequently they took a few migrant passengers on outward voyages, although never to the same extent as the vessels of Telaga Biru.

The Sapudi *lete-lete* were easily distinguishable from the Giligenting vessels by the lack of a 'doghouse' aft and were regarded as the most authentic examples of the *lete-lete* type. Some of the vessels built on Sapudi from the 1970s onwards were very large, the largest of all *lete-lete*. Following motorisation in the early 1980s, some of them became larger still, of a size to rival the largest Bugis vessels of that time. Yet although some vessel owners from Sapudi have been phenomenally successful, in general the Sapudi fleet has suffered from changing economic circumstances during the past two decades. The smaller-scale Sapudi operators, especially those from Nonggunong in the northwest of the island, have been less successful than the Giligenting owners in obtaining work in the timber trade. In particular, they lacked the capital to purchase and install engines. Moreover, being carriers rather than traders, entrepreneurs from Sapudi were little involved in the trading of timber in the ports of Java.

Consequently the number of vessels from Sapudi engaged in the timber trade has fallen considerably in recent years.

Current Prahu Types

Up until about 1980, the types of prahu used by the main maritime ethnic groups were quite distinct from one another and even served as ethnic boundary markers (Macknight, 1979; Hawkins, 1982; Horridge, 1981, 1986). After the installation of engines became standard in prahu fleets, the traditional types were discarded in favour of larger vessels with much imitation of modern vessel features and little or no concession to tradition. Nevertheless, these vessels are still constructed using traditional techniques, albeit with some use of power tools and without any formal plans. Moreover, many of these large motorised prahus still use the traditional lateral rudders. Very conspicuous and seemingly archaic, these are a convenient and efficient arrangement for shallow-draft sailing vessels, especially ones which routinely take the ground. However, they are inefficient for engine-powered vessels. For this reason many modern prahus with lateral rudders also have a central 'engine rudder' fitted to improve manoeuverability under engine power.

Although the Madurese prahus used to be the most distinctive and traditional of all sailing prahu types, the new-generation Madurese vessels are rather motley and uninspiring. Photo 21 shows one large type of about 300 cubic metres capacity, modelled after the large modern Bugis motorsailers but lacking their rakish style. The sailing rig is only token and sails are almost never hoisted. Photo 22 shows a smaller vessel of about 80 cubic metres capacity, which is only marginally efficient for the timber trade. This vessel is based on the type used by the Mandar sailors, characterised by highly raised topsides to increase carrying capacity. In this example, the hold is permanently open, to facilitate the loading and unloading of timber. This is obviously less safe than a full deck with a large secured hatch, especially when fully laden, but in practice the vessels cope satisfactorily due to the exaggerated freeboard. Most carry a crude sloop rig (one mast with mainsail and staysail or jib) as a standby in case of mechanical failure or to reduce rolling in beam seas. While this type of vessel has no connection with the traditional Madurese types, many other hybrid Madurese craft mix traditional bow and stern styles with incongruent superstructures and 'modern' sailing rigs. Finally, there are

Photo 21 A large Madurese motor vessel beached for maintenance at Giligenting, 2003. The lateral rudders have been temporarily removed but the central 'motor vessel' rudder, made from steel, can be seen.

Photo 22 A smaller Madurese vessel, about the minimum size viable in the timber trade in 2002. Unlike the vessel in Photo 21, the stern is sharp and the hold is undecked to facilitate loading and unloading of the cargo.

still a few engineless prahus from Madura working the timber trade, although they too have been modernised with sloop rigs and large aft superstructures for crew accommodation.

Regulations Governing Prahus

Prahus are officially classified as belonging to the Pelayaran Rakyat (People's Shipping). This classification dates back to the Netherlands Indies Shipping Act (*Scheepsvaart Verordening*) of 1935, which remained in force until 1964. The new Indonesian shipping law of that year retained the category of People's Shipping, thereby giving recognition to a transport service that hitherto had hardly received any official acknowledgment, despite the symbolic and strategic importance of prahus to the nationalist movement as well as the economic importance of prahu shipping following Independence (Dick, 1975a: 78–81). This formal recognition was furthered in 1971 with the establishment of Pepelra (Persatuan Pengusaha Pelayaran Rakyat, the Association of People's Shipping Entrepreneurs), nowadays usually abbreviated to Pelra. This organisation was set up to serve as a channel of communication between prahu operators and the government, with Pelra branch offices in all main ports.

People's Shipping is perhaps a more appropriate term than 'prahu shipping', because over the past 30 years these vessels have increased in size to the point where many are now well beyond the size range that could reasonably be classified as a prahu. In 2003 a typical 'small' prahu in the timber trade in 2003 was around 60 feet (18 metres) on deck by 20 feet (6 metres) beam, carrying about 100 cubic metres of timber, while the larger vessels were up to 100 feet (30 metres) long on deck with a carrying capacity of 500 cubic metres or more. In keeping with the increase in size of these vessels, the term *perahu* has been replaced in official classification terminology by *kapal* (ship). Nevertheless, the owners of these vessels and the men working on them still usually referred to them as *perahu*.

The classification of People's Shipping has economic advantages. Vessels in this category pay extremely low port dues by comparison with the rates charged for modern vessels, which come under the general classification of Pelayaran Nusantara (Archipelagic Inter-island Shipping). Prahus are also exempt from numerous restrictions applying to modern vessels. For example, prahu skippers do not require any formal qualifications. Whereas the master of a motor ship (*kapal motor*, KM) plying only Indonesian waters was required in 2003 to hold a minimum qualification of Restricted Shipping Master (*Mualim Pelayaran Terbatas*, MPT), awarded on completion of a three-month course at the maritime academy (Sekolah Akademi Kemaritiman) in Surabaya or Semarang, a prahu skipper required only a People's Shipping master's certificate (*Mualim Pelayaran Rakyat*, MPR), obtainable from the local harbourmaster with no study requirement. The fee for an MPR at the time was reportedly Rp 500,000. Similarly, a KM vessel had to have an engine room head with the qualification of *Mualim Diesel* (MD), again requiring formal study, whereas a motorised vessel falling under the People's Shipping regulations did not need any trained engine room staff.

To qualify for these exemptions, Rakyat vessels were required to be constructed of wood and to be propelled primarily by sail. However, small wooden motor prahus of less than 100 cubic metres capacity could also be registered as Rakyat vessels.[14] Above the 100 cubic metre limit, if an engine was fitted it was supposed to be just an auxiliary (*pesawat pembantu*), with the power of the engine restricted according to the size of the vessel. In practice, however, these and numerous other requirements were widely disregarded. The official size limits for People's Shipping have altered over the years in response to the size of vessels being used in the trade. In 1983 motor sailing vessels (*kapal layar motor*, KLM) coming under the Rakyat regulations could have a cargo capacity of up to 424.5 cubic metres. In 1986 this limit was raised to 850 cubic metres and in 2002, to cope with the increasing size of the largest Bugis motorsailers, it was increased to 1,415 cubic metres or 500 gross tons (Perla, 2002). However, most vessels in the Rakyat fleet were much smaller than this.

Although outside the time frame for this study, a further revision of the laws occurred in 2008, when the new Indonesian shipping law (Law 17/2008) scrapped the requirement for Rakyat vessels to carry sail. This was realistic because so many prahus had long ago dispensed with sailing rigs. Under the new law, to qualify for Pelayaran

---

[14] In practice, many prahus carry considerably more than their registered tonnages indicate because tonnages are often understated in order to minimise port dues (see Appendix 3). For this reason, in this book the capacity of vessels will be given in cubic metres and based on the actual volumes of timber that the vessels concerned have been known to carry.

Rakyat status, vessels had to have a capacity of no more than 500 cubic metres (175 gross tons) and be operated as "a people's enterprise with traditional features" (Dick, 2008). This definition refers to the 'non-corporate' organisational characteristics of prahu shipping, which are quite different from those of formal sector shipping (Dick, 1978). The later chapters of this book, focusing on the role of the Madurese in the Java Sea timber trade, will show just how different their approach to business has been by comparison with the mainstream corporate sector.

4

Opportunity and Illegality

Although the New Order government wasted no time in attracting foreign timber companies so that it could profit from the largely untapped forest resources of Kalimantan, it also recognised the need for increased reliance in Java for timber from the Outer Islands. Apart from the need to conserve Java's teak forests, both for hydrological purposes and the foreign exchange that teak could earn, timber was needed for the construction industry. Consequently, the rapid increase in the volume of timber landed in Java after 1966 accorded with the central government's plans for development. However, the manner in which this timber was being brought into Java did not meet with official approval. This was not just because most of the timber coming into Java was being carried by prahus rather than modern shipping as had been intended. Of much greater concern was that much of this prahu-carried timber was being smuggled, with no adequate documentation and widespread evasion of timber royalties to the central government.

While all of the prahu-operating groups were to some extent involved in this extra-legal timber trade, the Madurese have been the most conspicuous and the most persistent. This chapter outlines the nature of Madurese involvement in the Java Sea timber trade from the late 1960s until 2003, showing their main ports and routes, their specialist market niches and the strategies they employed to avoid official scrutiny as the government began to clamp down on undocumented timber shipments to Java. Particular attention is paid to the concept of legality in the timber trade and how the timber industry is controlled and revenues to the state assured through implementation of the timber transport regulations. Also shown are the consequences for Madurese prahu operators following the ethnic conflict centred in Sampit in 2001 and the introduction in the same year of regional autonomy, after which the timber-supplying districts of Central Kalimantan imposed substantial new taxes on timber being taken out. The last part of the chapter focuses on the ports of Sepulu and Telaga Biru on the north coast of Madura, both of which became conduits for extra-legal timber bound for the Java market during the 1990s. Both places were frequently targeted in 2002 by police in attempts to force compliance with the timber transport regulations, putting strains on their respective economies. However, because of differences in local leadership, one survived as a timber port while the other did not.

Taking Advantage of the Timber Boom

From the 1960s onwards, the Java timber market was supplied from South Kalimantan, mainly through the port of Banjarmasin, and South Sumatra. These two provinces were not as richly endowed in forest resources as East Kalimantan but they had the advantage of proximity to Java. East Kalimantan was unattractive for sailing prahus, not only because of the distance to Java but also because during the dry season, the main sailing time, the prevailing wind blows northward through the Makassar Strait, necessitating much tacking to windward in order to reach the Java Sea. A significant amount of timber for the domestic trade was carried from East Kalimantan by prahu but it went across the Strait to Sulawesi. By contrast, South Kalimantan and South Sumatra had distinct advantages for sailing prahus bound for Java.

Situated about 30 kilometres from the sea on the Martapura river, a tributary of the lower Barito, Banjarmasin has always been one of the most important ports of the archipelago. Its trade orientation was towards Surabaya rather than Jakarta because in either monsoon the passage could be made in both directions with the wind on the beam, enabling fast sailing and a more or less straight course on each leg. Voyage time in either direction was about three days. In practice, however, only one trip a month was normally possible because of the time spent in port at each end.[1] Some timber was also carried from

[1] For both motorised and sailing prahus, the time spent in port was more or less the same, usually about one week in Java and one to two weeks in Kalimantan. In either case all loading and unloading is usually done by manual labour. Larger prahus loading outward goods in Java usually spend longer in port in Java than smaller vessels, which sail empty to Kalimantan.

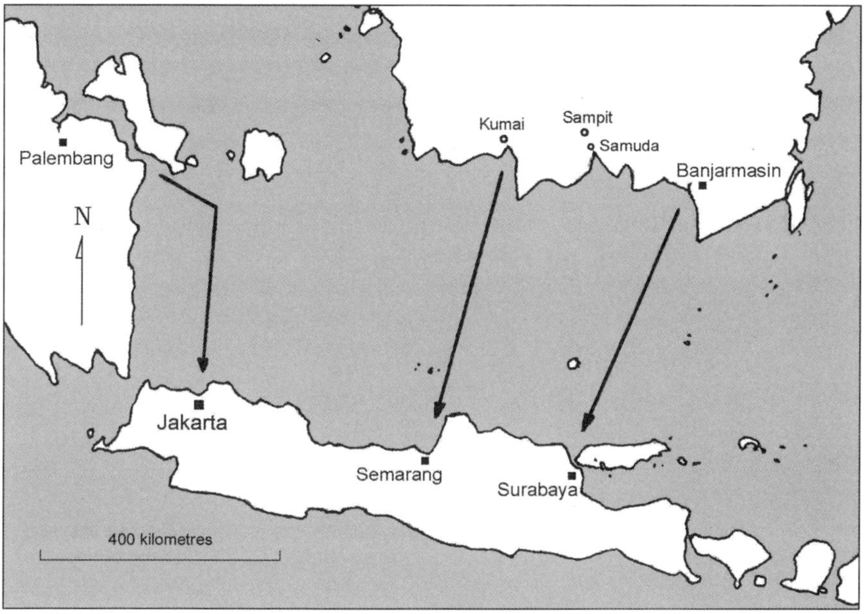

Map 10 The main routes for sailing prahus in the early phase of the timber boom

Banjarmasin to Jakarta but for sailing vessels this was a less convenient route. Although during the dry season the downwind trip to Jakarta could be done in just a few days, on the return trip sailing vessels had to tack along the north coast of Java as far as Madura before heading north. This was the most practical way to make distance to eastward because the long land mass of Java disturbs the prevailing wind for up to 50 kilometres out to sea. A diurnal 'seabreeze' sets in during the day as the warmer air rises over the land, enabling sailing vessels either to close with the land or to make distance eastward. Yet despite taking advantage of this diurnal wind pattern along the north coast of Java, the return passage alone took about three weeks. Consequently, the greater part of the timber shipped to Jakarta in the early years of the timber boom came not from South Kalimantan, but from the provinces of South Sumatra, including the large islands of Bangka and Beliton (which later became a separate province) and Jambi. The South Sumatra–Jakarta route is also largely on a north-south axis, enabling good advantage to be taken of the southeast monsoon in both directions. Sailing prahus bound for Jakarta usually had to tack to windward through the narrow Bangka Strait, but its sheltered waters are calmer than in the Java Sea and the passage time was usually only three or four days.

On these two principal timber routes, South Sumatra–Jakarta and Banjarmasin–Surabaya, prahus dominated the market because they could transport timber at rates well below those of modern ships. Small wooden motor vessels classified as *Lokal* (Local) ships were competitive with prahus,[2] but with their higher capital investment, they were less flexible in their movements and more dependent on carrying outward cargo from Java, with timber carried mainly as a return cargo.

Apart from Banjarmasin and the timber ports of the provinces of South Sumatra and Jambi, the other important supply area for the domestic timber market was Central Kalimantan. The main port for this province, Sampit, is well situated for crosswind sailing in both directions across the Java Sea, being almost directly north of Surabaya. However, it is inconvenient for engineless craft because it is situated about 70 kilometres up the Mentaya river. Partly for this reason and partly because motor vessels already enjoyed a substantial outward trade from Java to Sampit and needed a return cargo (Dick, 1975b: 99), prahus did not play a dominant role in the movement of timber from Sampit until they began to be fitted with motors from the late 1970s onwards.

Madurese prahus were particularly disadvantaged for reaching upriver ports such as Sampit because they could not tack through the eye of the wind as efficiently as could Bugis vessels with their European-derived rig. For example, to change tack on a *lete-lete*, the mainsail boom first had to be brailed up to the yard. The yard, boom and sail were then rotated forward around the mast, with one crewman pushing at the masthead and others heaving on a wooden lever temporarily inserted through the lower yard. Once the brailed up sail and spars had come around, the brail lines were released and

[2] The category of Lokal shipping applied to pure motor vessels (that is, not motor-sailers) of between 35 and 175 gross tons (100 and 500 cubic metres), of either wood or steel construction. Motor vessels below 35 gross tons came under the Pelayaran Rakyat regulations, while vessels larger than 175 gross tons came into the category of *Nusantara* (Inter-island) shipping. The distinction between Lokal and Nusantara shipping was dropped in 1988 with the creation of a single category of Domestic Shipping (Dick, 2008).

the sail reset on the new tack. This of course was slow, so that without anchoring, much distance could be lost to wind and current. Tacking on *golekan* and *janggolan* was more complex still, with the same net disadvantages.[3]

While they were capable of reaching Sampit, Madurese sailing prahus preferred instead to load timber at Samuda, just inside the entrance to the Mentaya and thus more accessible for engineless craft. Another convenient port for sailing prahus in Central Kalimantan and much used by both Madurese and Bugis prahus was Kumai, situated on a broad inlet well to the west of the Mentaya. From both Kumai and Samuda, the preferred destination was Semarang in Central Java, with the prevailing wind during the dry season again allowing good sailing times on both legs.

Writing of the situation in 1973, Howard Dick noted that timber transport was dominated by Bugis *pinisi* (*pinis*), large gaff-ketch rigged vessels each carrying between 100 and 200 cubic metres of timber. The Madurese prahus were considerably smaller with the commonest type, the *lete-lete*, having a typical capacity of 40 to 60 cubic metres of sawn timber. Overall, their role in the movement of timber was therefore minor by comparison with that of the Bugis, with many still involved in the carriage of salt and cattle from Madura and agricultural produce from Sumbawa or Lombok (Dick, 1975a: 96).

Nevertheless, from the late 1960s some Madurese vessels, both *lete-lete* and *janggolan*, had been involved in the carrying of timber from Kalimantan, combining this with the transport of salt to Jakarta and other places in the western part of the archipelago. However, unlike the Bugis, who carried full loads of sawn timber, the Madurese specialised in the transport of squared logs. This was a classic example of the Madurese approach to business, finding (or creating) a niche on the margin of the mainstream economy. There were two reasons for their specialising in the carriage of these large baulks of timber, which were more difficult to load than sawn planks. The first was that unlike the Bugis, whose cargoes were arranged in advance by forwarding agents, the Madurese lacked access to modern sawmills. In those days efficient mechanised sawmills were much less common than they were

[3] To describe tacking of these vessels in words alone is difficult. An excellent series of sketches showing the steps involved in tacking a *janggolan* can be found in Burningham and Mellefont (1997).

to become and many small timber traders simply set up hand-operated sawpits along riverbanks.[4] Sawing by hand — one man above, one or two below — was hard work and far less efficient than with modern machinery. Nevertheless, the trader could still make a good profit from his logs, which were in many cases illegally obtained, by simply sawing off the round outside part and selling the baulks. Second, these large baulks of wood were (and still are) in high demand in Java. The Bugis could not easily meet this demand because such non-standard items required direct negotiation with suppliers. The Madurese, with smaller vessels that could be deployed more flexibly, were thus able to monopolise the baulk timber trade. Moreover, whereas the Bugis vessels earned income from freight only, Madurese vessel owners who had the means became involved in the trading of timber. Large vessels make profits by virtue of their large freight capacity but small vessels can still be profitable if they can combine freight work with trading.

While in 1973 the Bugis were certainly dominant in the ports of Surabaya and Gresik, the role of the Madurese in the timber trade at this time may have been more significant than Dick implies. Hawkins, who carried out fieldwork in Indonesia in the late 1970s, noted that Madurese prahus were almost as common in the Java Sea as Bugis vessels; he also commented on the large numbers of Madurese prahus he saw in both Jakarta and Semarang (Hawkins, 1982: 41). In Jakarta, however, Bugis and Madurese vessels used different ports. The Bugis *pinis* (*pinisi*), with their lofty gaff topsail rigs, berthed in the river at Sunda Kelapa, conveniently located in the old city precinct. An important port since pre-European times, Sunda Kelapa (formerly known as Pasar Ikan) has in recent times become the premier prahu port of the archipelago as well as a magnet for tourists. Madurese prahus, however, discharged their cargoes at Kali Baru, a shallow back-water situated to the east of the container shipping terminal at Tanjung Priok and formerly used mainly by small fishing boats.

By the late 1970s Kali Baru was a teeming place, with the harbour routinely jammed during the sailing season with Madurese *lete-lete* and timber stacked high all around the quay. In the daytime, so many trucks were being loaded or making their way along the narrow street

[4] In 1983, when the author sailed on a Bugis vessel to Banjarmasin, there were still many pitsaw gantries set up along the river bank with logs being sawn into planks by hand.

Photo 23 A scene in Kali Baru harbour, 1984. The two large motor vessels in the centre are from Riau but all the sloping yards belong to *lete-lete* from Madura. Note the heavy 'laminated' yards in the foreground, built from three or more giant bamboo stems secured together with wire lashings every few inches.

that it was difficult to get through the throng even on foot. The volume of timber being discharged at Kali Baru may then have been as much if not more than at Sunda Kelapa, despite the 'showcase' status of the latter port and the smaller size of the Madurese vessels. For the year 1981, Hughes (1984) adjusted the official figures of timber landed in the two ports to fit with the reality of the vessels using them and concluded that Kali Baru was significantly the more important as a destination for timber.[5]

Of the prahus calling at Kali Baru, the majority were Madurese, mainly *lete-lete* from Giligenting, with lesser numbers of the same type from Sapudi, and *janggolan* from Sreseh. That is not to suggest that nearly all of the timber passing through Kali Baru was carried by Madurese prahus because motor vessels from Riau would have

[5] According to Hughes (1984: 94), 82 per cent of all timber landed in Jakarta in 1981 was carried by prahu, with 72 per cent of that discharged at Kali Baru.

carried a significant amount of the recorded total. But it does seem reasonable, from Hughes' adjusted figures, to assume that the total amount of timber carried by Madurese prahus into Jakarta at that time would have been close to, if not more than, the total carried by the Bugis fleet discharging at Sunda Kelapa.

As at Sunda Kelapa, most of the timber unloaded at Kali Baru was sawn in standard sizes. Nevertheless, the timber discharged at the two ports did not always have the same provenance, despite some overlap. Bugis vessels loaded their timber mainly in the main ports of the western archipelago: Palembang, Jambi and Pontianak, in that order of importance. Madurese vessels also loaded timber in those same ports, especially Palembang, but most obtained their cargoes in small remote inlets along the Java Sea coast of Sumatra: the Duabelas, Lambanghitam and Lumpur rivers in South Sumatra, and the Mesuji, Tulangbawang and Seputi rivers in Lampung (see Map 11). Madurese *lete-lete* and *janggolan* also carried much timber from the islands of Bangka and Beliton to Jakarta and Cirebon in West Java. The places where Madurese prahus obtained their timber were in most cases avoided by large Bugis vessels because they were too small to offer prospects for outward cargo from Java and too remote for the organisation of consignments by shipping agents. Prahus calling at these isolated inlets on the coast of southern Sumatra were still operating in a largely traditional manner. The skippers were less likely than in earlier times to be trading on their own behalf but they still arranged the cargoes by negotiating directly with owners of small sawmills along the banks of the rivers. Financial backing for the purchase of the timber was typically raised within the Madurese community. With fairly large amounts of money involved and no telecommunications readily available, trust was critically important. Skippers were usually given substantial bonuses to encourage loyalty to the vessel owners. The main departure from tradition was that there was now little scope for outward cargoes, the salt and livestock shipping trades from Madura having been taken over by either trucks or modern shipping. Unlike the large Bugis vessels, which carried freight on both legs of each voyage, Madurese prahus working the timber trade in most cases sailed empty from Jakarta. Profits from the timber trade were nevertheless sufficient for them to remain viable.

The concentration on sawn rather than baulk timber at Kali Baru was because the place was so crowded. The timber market at Kali Baru was immediately adjacent to the prahu harbour, with most

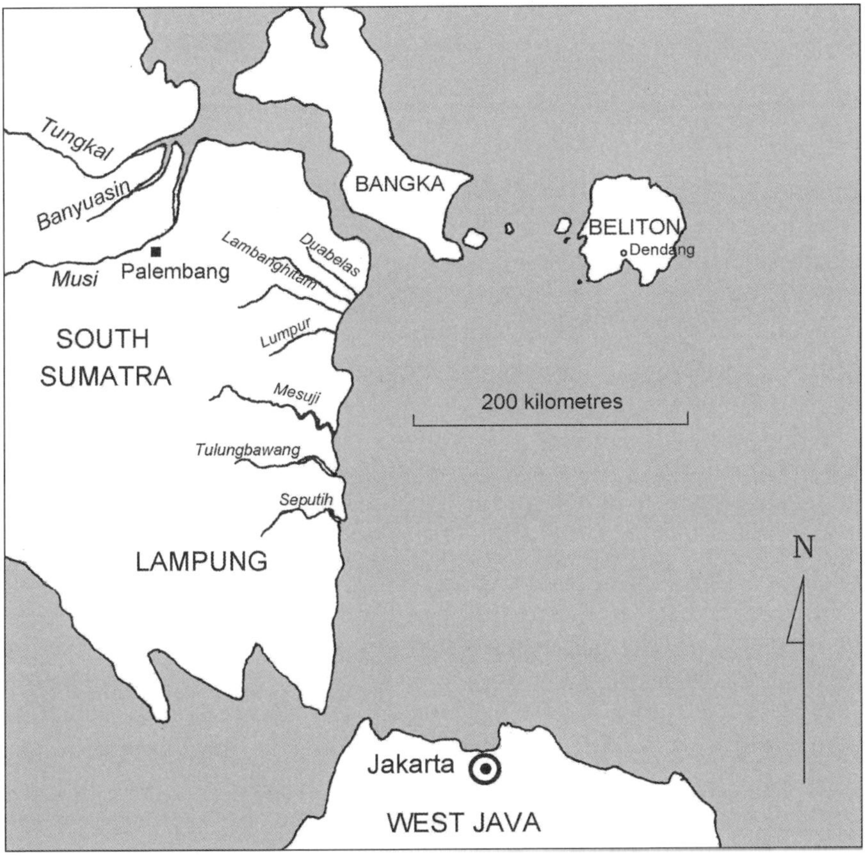

Map 11 Sources of timber in Lampung and South Sumatra for Madurese prahus

of the dockside covered in stacks of timber two or three metres high. Whereas stevedoring labourers would carry planks from the prahus and stack them directly on the quay, squared logs needed heavy lifting equipment for loading onto trucks, for carriage to warehouses. Baulk timber for Jakarta was therefore supplied mainly through Cirebon or Tegal, the latter a short distance across the border in Central Java.

In the case of *janggolan*, the largest of the traditional Madurese prahu types, involvement in the timber trade had started in the late 1960s with the carriage of baulk timber from Kalimantan to Central and West Java. In 1975, however, *janggolan* switched to Lampung, the southernmost province of Sumatra, to carry sawn timber from there to Kali Baru or Cirebon. Because these vessels were at that time still carrying salt from Sampang to Kali Baru, the move to Lampung made good economic and nautical sense. They obtained their timber mainly in the small port of Seputih, on the Seputih river, through a Madurese merchant named Haji Ismail, a native of Giligenting. Haji Ismail's business evidently prospered, for later he moved to Kali Baru. Many Madurese from prahu communities, especially Giligenting, became small-scale timber traders in West Java around this time. Although outsiders, they moved easily into this economic niche because in those days one did not need a large amount of capital to set up in the timber business. In the words of one former Madurese skipper, "with as little as ten cubic metres you could start up as a timber trader". With the national economy surging ahead under the New Order government, Jakarta was hungry for construction timber.

One example of extraordinary success at this time was Haji Sidik, a *janggolan* owner-skipper from the desa of Taman in the kecamatan of Sreseh (see Map 8). During his stays in Kali Baru harbour, Haji Sidik had noticed small motor ships coming into the port with cargoes of small poles, about 3 metres in length with a diameter of between 30 and 60 millimetres. Such poles, cut from saplings of mangrove, melaleuca and other species growing in swampy areas, were much in demand for scaffolding in Jakarta's booming construction industry. They were sold to pole traders at Kali Baru, who in turn supplied the building contractors. Haji Sidik learnt from these traders that the poles came from a place called Dendang, on the south coast of the island of Beliton (Map 11). In 1979 he determined to try his luck in this trade. With his only vessel, optimistically named *Harta Madura* (Wealth of Madura), he set sail for Beliton and found his way to Dendang, an isolated place with no township, at the head of a deep bay. There he arranged with a Chinese trader to purchase a load of poles. He sailed with this cargo of poles back to Kali Baru and sold it for a handsome profit. Thereafter Haji Sidik continued to work the pole trade from Dendang to Kali Baru, and also to Cirebon. He prospered and became by far the wealthiest person in Sreseh. As his wealth grew, he ordered one new *janggolan* to be built after another, until he owned 15 large prahus, a huge personal fleet in this traditional maritime milieu. All his new vessels were built to the traditional design, including the indigenous single-fluked wooden anchor of the same type as had been used by large *jong* four centuries earlier (Photo 24). None had an engine, even though engines were by that

Photo 24 Hauling up the traditional wooden anchor on a *janggolan*, from a depth of about 50 metres, 1986. The single fluke is completely covered with dense mud, proof of the efficiency of this anchor of indigenous design.

Photo 25 A *janggolan* laden with poles from Beliton, 1984. The interior is filled almost to the deckhouse roof. These prahus have two rudders but use only one at a time, always on the leeward side with the other stowed along the deck, hanging over the stern. The rudders are six metres long and made from dense timber. When tacking, their position must be reversed, a difficult task given their size and weight.

Photo 26 Empty *janggolan* moored together off Kali Baru, 1984. On these prahus the yards and their support poles remain standing in port, while the sails are slid down on rings and furled. The sailors do not normally need to do any work aloft.

time standard among Bugis vessels. They were also substantial craft, carrying up to 200 cubic metres of sawn timber or 10,000 poles. Other Sreseh owners followed Haji Sidik into the pole trade and they too prospered, with the profits better than from carrying sawn timber. Faced with this intense competition, the motor vessels that previously had this lucrative niche to themselves withdrew, unable to match the landed price of poles carried by the cheaply operated prahus. With their higher capital outlay and operating expenses, the motor vessels could not afford to wait in the 'queue' for their turn to load. Thereafter vessels from Sreseh dominated the pole trade from Dendang to West Java and some were still working it in 2003.[6]

While *janggolan* were well represented at Kali Baru, most Madurese vessels using the port were from Giligenting. So profitable was their business of transporting and trading timber from South Sumatra that many prahu owners from Giligenting relocated to Kali Baru during the sailing season to monitor the operation of their businesses. They bought or rented houses in the densely settled kampung around the harbour. Some vessel owners from Sreseh also set up households in Kali Baru or Cirebon. These vessel owners would typically stay in West Java throughout the sailing season, returning to their home villages and primary households with the onset of the northwest monsoon. For economy, they always travelled by bus. Their wives usually remained at home in Madura, although in many cases an adult daughter or perhaps a daughter-in-law married to a prahu skipper would stay with them in West Java as a housekeeper. Some vessel owners even took on second wives in West Java, thereafter maintaining two complete households. For the prahu sailors, however, such comforts were financially out of reach. While in port at Kali Baru or Cirebon, a skipper might stay with his employer, who very often was a relative anyway, but most crew members would sleep on board their prahus.

Accommodation standards on these traditional vessels were very simple, with everyone sleeping together on the hard deck of a small after cabin bulkheaded off from the hold. Prahu sailors were used to such conditions, however, and they always looked forward to their stays in Kali Baru. They usually had plenty of spare time in port

[6] A few Butonese *lambo* also carried poles from Beliton to Kali Baru but they obtained their poles from Membalong, in the southwest of the island.

Photo 27 A *lete-lete* from Giligenting with a cargo of sawn timber from South Sumatra, bound for Jakarta, 1984. Tacking is usually done with the wind from astern: after furling the forward and after sails, the large mainsail is allowed to run freely around the front of the mast before being sheeted home on the other side. In the process the yard rotates over the top of the masthead.

Photo 28 Loaded to the limit with timber, a *lete-lete* from Sapudi ghosts towards the West Java shore, 1984.

because the hard labour of unloading the timber was carried out by stevedoring workers. While their seafaring vocation provided them with a source of income, it was also in keeping with their tradition of experiencing the world beyond their home villages, a tradition that had developed as a consequence of their outward-oriented economy. Kali Baru was the best place of all, with so many fellow Madurese there, so many friends to catch up with, so many prahus and houses to visit, the crowded market place with its numerous food stalls, and of course the local bordello kampung. As well as the camaraderie, there was a sense of pride among the Madurese sailors in this teeming port in the nation's capital. Here, as outsiders, they were 'in'. Although small engineless cargo vessels had been eclipsed throughout the rest of the world, and even within much of Indonesia, these were halcyon times for the seafarers and maritime entrepreneurs of Madura.

Some of the most successful Madurese prahu owners and timber traders even established their own prahu shipping agencies in Kali Baru, considerably increasing their opportunities for profit. Until 1964 individual prahu skippers could deal directly with port authorities, but in that year a law (No. 5 of 1964) was introduced, stipulating that port business for prahus could be carried out only through licensed shipping companies owning a minimum of two vessels (Dick, 1975a: 81). The intention of this was to streamline port administration, as implied by the term for shipping company: *perusahaan expedisi* (expediting company). Although such companies are regarded for port administrative purposes as the 'owners' of the vessels they represent, the vessels concerned are usually owned by other individuals, who are charged a rate by the shipping companies for services rendered. To set up an expediting agency requires considerable capital as well as permission from the authorities (Directorate-General of Sea Communications) and the local Pelra office, with such permission given only if a need exists. Such a need certainly existed at Kali Baru from the late 1970s onwards. With around 2,000 calls each year to that port by Madurese prahus, it was appropriate that there were some expediting companies there headed by or at least staffed by Madurese, just as at Sunda Kelapa the shipping companies were operated by Bugis or Makassarese.[7] One such Madurese prahu owner who established

[7] Hughes (1986: 106) notes 2,535 calls made by sailing prahus at Kali Baru in 1981. The great majority of these would have been Madurese vessels.

a *perusahaan expedisi* around this time and became very successful was Haji Hosin, of Aeng Anyar, Giligenting, whose company, P.T. Sendi Gita Pala, still had an office in Kali Baru in 2003. Haji Hosin had settled permanently in West Java but continued to return each year to Giligenting for the Lebaran celebration at the end of the fasting month.

The Legality of Timber Shipments

Although the success of these Madurese prahu operators throughout the 1970s and 1980s depended in most cases entirely on the booming timber trade, most of the timber coming into Java was illegal. This was reflected in the great discrepancy between the official figures for timber carried by prahus and the reality. For example, the data presented by Dick for Surabaya in 1972 indicates that the official figure for the volume of timber landed at that port was slightly more than one-sixth of the actual amount landed, with the remaining five-sixths being undeclared and thus illegal (Dick, 1975a: 95). Similarly, for Jakarta in 1981, Hughes estimated that the official record of the volume of timber landed in Sunda Kelapa represented only about one-third of the actual amount landed, while the official figure for Kali Baru, the port for Madurese prahus, was only one-tenth of the actual amount landed (Hughes, 1986: 107)

Neither of these authors explains the factors behind such massive understatement of timber cargoes, except to note that understatement of prahu cargoes in general is a common practice in order to reduce the amount of taxes payable. Dick does, however, give the example of copra, for which certain taxes were supposed to be paid in the port of loading. Apparently many prahu shipments of copra evaded these taxes altogether until 1972, when the Copra Board set up 'prahu posts' in East Java, to charge levies upon cargoes without a certificate of payment of levies from the port of loading.[8] This reform could have been disastrous for the prahus in this trade but in the event the impact was fairly low because the prahus typically reported only 20 per cent of the cargo (Dick, 1975a: 97).

The rationale for the under-reporting of timber cargoes was more complex. Timber for commercial purposes is supposed to be obtained

[8] Copra-carrying prahus, mainly from Buton, discharged their cargoes in Gresik.

from areas that have been approved for logging, with the logging being carried out by the holder of the concession or permit. All other timber from designated forest areas, except that taken by indigenous people for their own use, is *ipso facto* illegal. Moreover, as the 'owner' of the resource, the state expects a royalty on wood taken. Because of the vastness and isolation of forest areas, however, timber extraction has always been difficult to monitor and control. To get around this problem, the Dutch implemented a policy much like the pass system they had devised in the seventeenth century to control indigenous shipping. Just as every ship, no matter how small, was required to obtain a pass for every voyage — a requirement that persists until the present — so too every shipment of timber was required from 1932 onwards, under the *Bos Verordening Java en Madoera* (Timber Ordinance for Java and Madura) of that year, to have its own special transport pass. By focusing on the movement of the commodity from one administrative district to another rather than its extraction from the forest, a high degree of control became possible at minimal expense.

The 1932 Timber Law was aimed primarily at the movement of teak within Java and Madura, reflecting the colonial state's preoccupation with that timber. It endures in principle in the Indonesian state through Law 64 of 1957, on the basis of which regional administrations issued their own regulations requiring that all timber consignments have an official transport pass, obtainable from the district forestry office. Like its colonial predecessor, this pass applies to movement of any consignment of timber, whether by sea or land, to a place outside the administrative district (kabupaten) of origin. The official term for this document used to be the *Surat Angkutan Kayu Olahan* (transport permit for processed timber), widely referred to as SAKO. Since 1999 the SAKO has been replaced by the *Surat Keterangan Sahnya Hasil Hutan* (SKSHH, certificate of legality of forest produce), a blanket document that can also apply to non-timber forest products such as rattan. Any consignment of timber arriving in Java without such a document is illegal.

In order for the shipment to qualify for an SKSHH (or SAKO) and thereby become legal, the owner of the timber — either the logger (logging syndicate or company) or the sawmiller who has subsequently obtained the logs — must pay the forest produce royalty fee, formerly the *Iuran Hasil Hutan* (IHH, forest products levy), now the *Provisi Sumber Daya Hutan* (PSDH, provision for forest resources). This royalty fee, the main tax on timber, is calculated on a per cubic metre basis. It varies across species with the most common timber, meranti, being taxed at the lowest rate and high-demand timbers being taxed at higher rates. A smaller reforestation fee (*Dana Reaboisasi*, DR) is also payable at this stage. These fees are typically paid for by the purchaser, although for official purposes the seller of the timber may be deemed as the payer. The transaction with the district forestry office is usually carried out by either the seller or, in larger ports, the shipping agent, with reimbursement from the buyer in either case.

In order to prevent deceptions such as a permit for a small shipment being used for a large one, each timber transport permit includes an attachment stating the number of lengths of timber and their sizes, as well as the species, of the shipment concerned. This is not the same as the vessel's manifest, which is attached to the port clearance (*surat izin berlayar*). The manifest states simply that a cargo of timber is being carried, often without details, and it has no validity for port forestry officials. Upon arrival in the destination port, the SAKO/SKSHH can be shown as proof that the necessary taxes have been paid and that the cargo is therefore legal. If there is no such document, or if the description on the attachment does not match the timber carried, the entire cargo may be confiscated and auctioned.

The above description outlines how the transportation of timber is supposed to be controlled. In practice, while the requirement for a transport permit was taken seriously in Java from the 1950s onwards as a means of controlling the pilfering of teak, it was not rigorously enforced elsewhere. In so far as shipments by Madurese prahus were concerned, the relevant regulations might just as well have not existed during the first two decades of the boom in the Java Sea timber trade.

In 1974 the Directorate of Timber Marketing expressed its concern over the lack of application of the law concerning the movement of timber:

> The legal framework thus already exists; but of late, especially with the increase in inter-insular trade, implementation of the regulations still leaves a great deal to be desired. On any given day one can still see numerous shipments of timber without an official stamp and transportation pass, no matter that these are obligatory... (Kehutanan, 1974, B: 13.)

It is possible that the timber being carried by the Bugis into Sunda Kelapa did show a modicum of compliance with these official requirements because their cargoes were organised through forwarding agents. In the case of the movement of timber by Madurese vessels, the view expressed above would seem to understate the real situation. Skippers of Madurese prahus negotiated for cargoes directly with operators of small sawmills — including many operated by hand using pit-saws — along the banks of the rivers of southern Sumatra and the southern rim of Kalimantan. Many skippers traded on their own behalf, while in the case of those buying on behalf of timber merchants, the merchants were also Madurese, based in West Java. According to Madurese shipping agents interviewed during fieldwork, virtually none of the timber carried by Madurese prahus came from HPH concession areas allotted by the central government. That did not necessarily mean that the timber had been illegally obtained, because much of it may have come from small concessions of up to 100 hectares parcelled out at the district level. Nevertheless, the timber imported to Java by Madurese prahus during the 1970s and 1980s was illegal in the sense that no royalties were paid and no transport permit obtained. This was no minor issue or mere technicality because the revenues payable to the state were large (see Chapter 5).

Every former skipper, agent or timber trader from Madura with whom I spoke was adamant that regulations pertaining to the SAKO were not applied until about 1990 or even later, depending on the port concerned. In their experience, the SAKO had not applied to them hitherto. Had there been any awareness of a law requiring a transport permit and substantial taxes to be paid in order to qualify for it, their understanding of it would have been vague at best, with any concerns allayed by a notion that it did not really apply to Rakyat vessels.

The illegality of timber transported by prahu underlay the massive discrepancies between official figures for timber discharged in port and the reality. In dealing with the port authorities, an agent would declare the volume of the shipment according to the figure stated on the SAKO, which might be perhaps one-third of the actual amount carried. Possibly the Bugis vessels calling at Sunda Kelapa in 1981, with an average recorded cargo of only 99 cubic metres of timber despite an average capacity nearly three times as great, were carrying such partially legal cargoes (Hughes, 1986: 107). In the event that no SAKO was held, as was often the case, an agent would enter the lowest volume that would not arouse suspicion, to minimise the local taxes that were payable in ports in Java on each cubic metre of timber landed. The volume of timber carried was rarely checked physically because it would involve unloading and stacking the entire cargo for inspection.

Senior forestry officials in Jakarta wanted to change this state of affairs. As early as 1974, the Jakarta Forestry Service was expressing concern over the frequent absence of any timber transport permit at all for shipments from the 'Three Rivers' area (the Musi, Banyuasin and Tungkal estuaries near Palembang, shown on Map 11). It also frequently encountered SAKO permits from both Lampung and Bangka without the appropriate stamp from the district forestry office, with one such case under prosecution at the time (Kehutanan, 1974, E: 2). It is not known whether the vessels concerned were Madurese or Bugis. In any case, such prosecutions seem to have been very much the exception, perhaps to set an example rather than to attempt to deal head-on with such widespread flouting of the law. Not only was there little serious attempt on the part of regional forestry officers to enforce the law in ports of loading, but the efforts of port forestry officers in Jakarta seem to have been too spasmodic to achieve any curbing of illegal timber traffic prior to the mid-1990s.

From the perspective of citizens of more developed countries, such a widespread disregard for the law and evasion of revenues payable may seem outrageous, especially when coupled with the environmental implications. However, the issue of illegality in the Java Sea timber trade needs to be considered in context. In the first place, the Madurese were not the only group involved in illegal timber imports to Java, for Bugis and Mandar vessels also routinely carried illegal timber cargoes. Second, the popular notion among the Javanese that the world of the prahu abounds with nefarious types with a bent for illegal dealings is misleading. Among the Madurese, some of the skipper-traders I have known have been men of genuine piety who would regard it sinful to cheat anyone. Yet these same men, pillars of their society, continued routinely and deliberately to contravene the timber transport regulations and to cheat the state of large amounts of revenue, even at a time when vigorous attempts were being made to stamp out such practices with heavy fines and prison sentences being meted out to offenders. Third, the traders, transporters and officials involved in this illegal timber trade did not necessarily represent an aberrant group in a wider law-abiding society. Rather, illegality in general, and corruption in particular, remain astonishingly pervasive in Indonesia, as is made clear in a new and revealing book on the subject (Aspinall and van Klinken, 2010). This is not because of an absence of appropriate legal frameworks but rather because the laws which are so commonly and flagrantly broken do not reflect deeply held societal values. While crimes such as robbery and murder are certainly not tolerated, corrupt dealings are regarded as normal, seemingly what most people would do if they had the opportunity or found themselves in a predicament that could be overcome by such means. The low salaries of government workers have encouraged the persistence of this corruption, with officials in virtually every field taking some opportunity to supplement their salaries with bribes. Consequently, it is very often in their own interests, and sometimes even in the broader public interest, to allow a degree of non-compliance.

It is therefore hardly surprising that in the case of the Java Sea timber trade, the many actors involved were reluctant to accept and comply with the laws governing the movement of timber. The word used in Indonesia to refer to the process of gradual acceptance of new laws is *socialisasi* (socialisation), an appropriate term because it implies acknowledgement of the cultural dimensions of the problem: that the state is not integral with the social world of those involved, and that much public education may be needed to achieve compliance with laws that do not correspond to prevailing cultural values.

The Decline of Madurese Involvement at Kali Baru

Although at Kali Baru completely illegal cargoes of timber remained the norm into the 1990s, port forestry officials (*Polsus Kehutanan*, special forestry police) had gradually been increasing the pressure on timber traders and prahu skippers to comply with the regulations on the movement of timber. These efforts were concentrated in the main ports: Jakarta, Surabaya and, to a lesser extent, Semarang. In Jakarta pressure was most effectively brought to bear at Sunda Kelapa, perhaps because it was a government port, with entrance gates, agents leasing premises for their offices and warehouses, and no resident community. Kali Baru, by contrast, was densely populated with a large fishing village located within the port precinct. There was no entrance gate and no clear separation of the community from the port proper; shipping agents operated from houses rather than separate office buildings. It was thus a 'community port' and most local households benefited in some way from the timber traffic. In this situation, the law was more difficult to uphold because it was easy to organise networks of collusion and corruption embracing shipping agents, Pelra officials, staff of the harbourmaster's office (*kesyahbandaran*) and police. There was also a real possibility of violent retribution towards any official who might become over-zealous (Dick. 1975a: 89).

Nevertheless, during the mid-1990s checks at Kali Baru by special teams (*tim*) became increasingly frequent, with heavy penalties imposed for timber without documentation and occasional confiscations of cargo. In such cases the burden fell upon the buyer of the timber, who might be the owner-skipper of the vessel but was more likely to be a local timber trader. Many of the shipping agents were also involved to some extent in trading, providing finance to pay for timber shipments.

The owners of the large motor vessels calling at Kali Baru, most of whom were Chinese, responded to this increased official vigilance by obtaining a SAKO for each shipment of timber, albeit not for the entire cargo. This naturally reduced the earnings for them and their financial backers but their vessels were large enough to produce a profit even after paying the relevant taxes. The smaller Madurese vessels lacked this economy of scale and could not remain competitive if a SAKO had to be obtained. Some Madurese timber traders at Kali Baru went bankrupt around this time, while many others were forced to withdraw from the timber business, turning to the small food stall (*warung*) trade in Jakarta, specialising in barbecued skewered meat pieces (*sate*).

Meanwhile, as more and more large motor vessels began to use Kali Baru, port costs increased significantly. The large Chinese-owned motor vessels were not Pelayaran Rakyat vessels, and were instead registered as motor ships (*kapal motor*, KM). Operating as always at maximum capacity, the port effectively became a port for motor shipping only. By the early 1990s many of the Madurese vessels using Kali Baru, and certainly all of the new and larger vessels among them, were also motorised. They were still registered as Rakyat vessels, which qualified them for a discount berthing rate. However, because many of them had no sailing rig as required by the regulations and had capacities of well over 100 cubic metres, the upper limit for motor vessels in the Pelayaran Rakyat, they should really have been registered as motor ships. The Madurese owners did not want to convert their vessels to KM status, for good reason. First, such a move would have eroded their profitability because port charges for motor ships were much higher than for Rakyat vessels. Second, motor ships were subject to more stringent safety regulations, whereas most of the Madurese vessels using Kali Baru did not even carry a radio. Third, motor ships were required to have a formally qualified master and engineer. While the study required for these qualifications was not onerous, no Madurese prahu skipper was so trained. Indeed, many had little schooling beyond elementary level.

Although the harbourmaster's office at Kali Baru may not have been concerned about these large Madurese motor prahus being registered as Rakyat vessels, it was no longer prepared to give them the discount berthing rate in the crowded port. If Madurese motor vessels were to continue to discharge timber there, they would have to pay the same rate as the KM vessels. Because this was unacceptable to the owners, Madurese motor vessels ceased to call at Kali Baru. As the Madurese presence in the port dwindled, the engineless prahus also withdrew, so that since the mid-1990s Kali Baru has been used only by small wooden motor ships owned by Chinese or Bugis entrepreneurs. In 2003 these motor ships were each carrying on average about 800 cubic metres of timber, around 10 times that of the typical traditional prahu from Sumenep sailing in the late 1980s. Kali Baru is still more important than Sunda Kelapa as a timber port, although unlike Sunda Kelapa there is still very little outward cargo.

Nevertheless, the move away from Kali Baru did not seriously damage the Madurese prahu economy. The large *janggolan* in the Beliton pole trade, all of which were still engineless, moved to Cirebon, while the Sumenep prahus moved their operational bases to Central Java, notably Tegal and Juwana. Madurese prahus had long been familiar callers at these two Central Java ports as well as at Semarang; now there were just more of them. Forestry department officials did make occasional appearances in these ports but in general things went smoothly, despite most of the timber landed being illegal.

With the shift of sawn timber imports by Madurese vessels from West to Central Java, it no longer made economic sense for these vessels to obtain sawn timber from Sumatra. Instead, they turned to Central Kalimantan, where the supplies were more plentiful. Moreover, for those Madurese prahus still relying on sail alone, the Central Kalimantan–Central Java route was convenient with the wind more or less on the beam on each leg, whereas to sail from Sumatra to Central Java would have involved much tacking with extra time and distance.

By the mid-1990s the traditional sailing fleet had shrunk considerably. Although in Sreseh loyalty to the *janggolan* remained absolute, at Giligenting all new vessels built from the early 1980s onwards were pure motor vessels, while most of the *lete-lete* had been converted to *sepel* with sloop rigs that made them much easier to tack. Many of these former *lete-lete* had also been fitted with engines. This did not mean that weather considerations became irrelevant, because in many cases the engines were too small to drive the prahus well into a rough head sea. This problem was compounded by the *lete-lete* hull shape, which was fairly full forward. For sailing cargo vessels such a form was not necessarily a handicap because the greater resistance of the blunt bow was offset by greater ability to withstand the pressure of wind on the sails, thereby producing more driving power. For pure motor vessels, however, fullness forward can be very detrimental to performance, especially when going directly into a heavy head sea. Apart from considerations of passage-making, having an engine did make a really significant difference when it came to making way upriver in Kalimantan. For medium to small vessels, this was the most important reason for the fitting of an engine. Although sailing prahus could with patience be worked into some remarkably awkward places, using poles and anchoring when the current was adverse, often two or three tides would be needed before a destination could be reached. For ports well upriver, most skippers of engineless prahus preferred to pay for a tow from a local motorboat. Because this was quite expensive, most owners of sailing prahus fitted an engine if they could afford to do so. Many took on partners who supplied the engine, a matter discussed in Chapter 5.

Revenue Evasion in East Java

The prahu port for Surabaya is situated at Tanjung Perak on a small river known as the Kali Mas, adjacent to the Madura ferry terminal. This busy port has long been dominated by Bugis and Makassarese vessels and since about 1980 they have been almost the only vessels to berth there. In the early 1970s Madurese prahus also used to call at the Kali Mas but they later ceased to do so because of the congestion, with preference given to the larger Bugis prahus.

The Madurese prahus that had been using Tanjung Perak relocated in three directions. First, some Madurese owners arranged for their vessels to discharge timber at nearby Gresik, which being only about 20 kilometres away by road was still conveniently close to the main market of Surabaya. Nevertheless Gresik had its disadvantages.

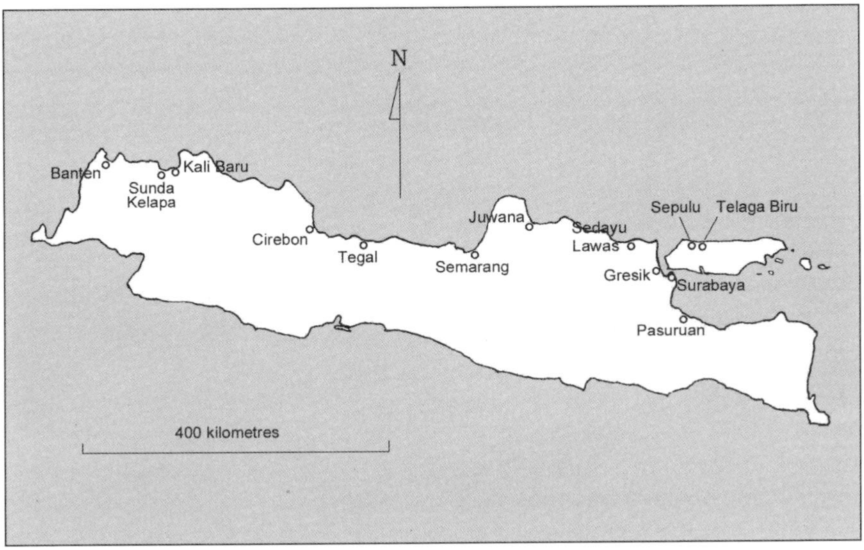

Map 12 Main timber destination ports of Java and Madura

Because the prahu harbour was very busy with many Butonese *lambo* calling to discharge copra from eastern Indonesia, there was little docking space and cargoes had to be taken ashore by small boats.[9]

A second option, taken by many Madurese prahu owners, was to discharge timber at the small inlet of Sedayu Lawas, a rather isolated place on the north coast, not far to the west of Gresik. This had once been a busy place for prahus but with the development of land communications and transport networks it had become a backwater. By comparison with any of the other places mentioned so far, Sedayu Lawas hardly rated as a port during the middle decades of the twentieth century, with no harbourmaster or port administration and no modern facilities. Nevertheless, for small-scale Madurese prahu operators it offered distinct advantages for the landing of timber from Kalimantan. Because there was no administration, there were no port charges. Shipping agents were neither available nor needed, further reducing costs. It was easy for sailing vessels to approach and enter and with plenty of room to anchor, and convenient for maintenance

[9] In the 1990s the prahu harbour at Gresik was upgraded to become the most modern and well-organised prahu port in Java.

Photo 29 A *beluntu* from the island of Bawean, at Sedayu Lawas, 1986. These prahus from Bawean were working alongside Madurese prahus in the illegal timber trade at Sedayu Lawas. The hull shape is derived from European vessels, while the rig is similar to that of the *kacik* (see Photo 6).

because vessels could dry out on the mud at low tide, with a nearby schoolyard often used for making sails. Most importantly, being situated a long distance by road from Surabaya and off the main north coast highway, it was not visited by forestry officials.[10] On the several occasions I visited Sedayu Lawas during the 1980s there were 20 to 30 prahus there, mainly *lete-lete* and *janggolan*, as well as a few *cemplon* from Lamongan and *beluntu* from Bawean. All were engineless. In effect it was a Madurese haven, with timber the sole commodity passing through. Needless to say, no one bothered to obtain a SAKO. Most of the timber landed there was quickly trucked away to the nearby towns of Tuban, Lamongan and Bojonegoro.

[10] The north coast highway from Surabaya meets the coast at Tuban, about ten kilometres to the west of Sedayu Lawas. The coastal road between Tuban and Gresik is in general used only by local people, who in the 1980s still had to rely on *dokar* (pony carts) for transport along part of this road.

The third option for Madurese prahus that had formerly discharged timber at Tanjung Perak was to relocate to Pasuruan, to the east of Surabaya. This had long been an important port for Madurese vessels and had the advantage of being a predominantly Madurese-speaking area. However, by comparison with Sedayu Lawas it took an extra day for sailing prahus to reach Pasuruan. Coming from South Kalimantan, the usual approach was via the eastern end of Madura rather than the narrows of the Madura Strait. This avoided the crowded Surabaya roads and for sailing vessels also allowed the best chance of using a fair wind to reach port. Being very narrow, the inlet at Pasuruan was also more difficult than Sedayu Lawas to enter under sail. Nevertheless, the prahu sailors were expert at ghosting in under minimal sail area, keeping speed as slow as possible so that they could manoeuvre the heavily laden vessels with bamboo poles. In the case of *lete-lete* and *janggolan*, the mainsail was usually furled at the entrance and the port entered under foresail alone.

Apart from Madurese prahus, Pasuruan was the main port in Java for Mandar vessels sailing from the east coast of South Kalimantan. It was also much used by prahus from Bima (Sumbawa), operated by Bugis who had settled in that area, as well as by prahus from Makassar. It is a true prahu port, because the river on which it is situated is too small and shallow for ships to enter. Indeed, at low tide many prahus lie on the mud. Until the 1980s deepsea ships still called at Pasuruan to load tobacco but they had to anchor offshore and rely on lighters. Although it has always been a busy port, serving a large town and a densely populated hinterland, it was slow to develop as a timber port and for the first two decades of the timber boom, was much overshadowed by Surabaya. However, being only about an hour and a half by road from both Surabaya and the large inland city of Malang, Pasuruan was poised in the late 1980s to become an important conduit for timber from Kalimantan.

After 1990, when forestry officials and police became strict about enforcing the law on the documentation of timber arriving in Surabaya and Gresik, those two ports became less attractive to smaller prahus. Sedayu Lawas was one alternative but Pasuruan was much the more important port and closer to the main market. Consequently, once Madurese prahus had been fitted with engines they tended to go to Pasuruan. Possibly because it had become purely a Pelayaran Rakyat port, a fairly relaxed attitude still prevailed there among port officials.

For example, to minimise port dues many prahus calling at that port claimed ludicrously low gross tonnages, shown on their operating permits (*pas kecil*) (see Appendix 3). Such fictitious gross tonnages would have been unacceptable in West and Central Java. Similarly, most prahus had no documentation for their cargoes of timber. Nevertheless, strictly speaking it was not the responsibility of port administration staff to police the timber transport regulations. This was a task for the police and port forestry officials, who in Pasuruan paid little attention to the matter of timber documentation until the mid-1990s.

Because vessels calling at Pasuruan were routinely discharging cargoes for which no SAKO had been obtained, the landed cost of timber there was significantly cheaper than at Surabaya or Gresik, while the cost of transhipping by truck from Pasuruan to Surabaya was not much more than from Tanjung Perak or Gresik. Thus many consumers in Surabaya, and especially in Malang, switched to ordering their bulk supplies from timber merchants in Pasuruan. As a result, Pasuruan became more important as a timber port and more and more timber traders set up warehouses there.

By 2003 there were about 50 timber trading businesses in Pasuruan, each with its own warehouse. Most of the warehouses were within a few blocks of the port area. As might be expected for a place with so strong a Madurese presence, the majority of these traders, including many of the most successful, were Madurese. The remainder, about 30 per cent, were mostly Banjarese or Bugis. Some timber trading businesses were being supplied by only one prahu, while the most successful ones were being supplied by ten or more prahus. The prahus concerned could be owned by the timber trader but in most cases they were owned by outsiders.

At about the same time as Pasuruan was emerging as a conduit for timber from Kalimantan, increasing amounts of timber began to be landed also at Telaga Biru on the north coast of Madura. Small amounts of timber had long been landed there as return cargo on outward voyages carrying passengers and cattle, but only for local consumption. Since the late 1970s, however, a few large motor prahus of non-traditional model from Telaga Biru had been carrying full shipments of timber from West Kalimantan to Surabaya. In the late 1980s, after the forestry department authorities in Surabaya had started to demand that a SAKO be obtained for every shipment, it occurred to these entrepreneurs from Telaga Biru that they could land their cargoes in their home port rather than Surabaya. Because there were no forestry officials in Telaga Biru, they could avoid having to pay for a SAKO, with the landed price for the timber being as a result much cheaper than on the Surabaya market. For prospective buyers in Java, the extra transport costs and time required for the vehicular ferry would be minor considerations when set against the lower cost of each truckload. This scheme received full approval from the desa head, a former prahu sailor. He was well aware that it would be illegal, in fact timber smuggling, but he recognised the potential for the local economy. Besides, who was going to pay attention in such a backwater as Telaga Biru?

This move was a success and soon all the local vessel owners became involved in the timber trade. The Telaga Biru traders specialised in high-quality timbers such as *bangkirai* or *benuas* (*shorea laevifolia*; *shorea laevis*), which they obtained from West Kalimantan. Bangkirai is a superb timber, dense, strong and durable with good workability despite its hardness, and much in demand for making items such as window and door frames and any high-grade construction work. With the market flooded with meranti and kapur timber, much of it of indifferent quality, these entrepreneurs of northwest Madura were able to exploit the niche for high quality timber, offering it at rates well below those prevailing in Java.

After 1990, business boomed in Telaga Biru. A continuous row of warehouses was built along the harbour foreshore and lorries stacked with sawn timber from Telaga Biru became a common sight in the vehicular ferry queue at Kamal. Most buyers were from Surabaya but some were from as far afield as Bali. The local officials at Telaga Biru, especially the staff in the harbourmaster's office, were naturally aware that the large quantities of timber that were being landed and subsequently sold to Java were illegal, but no one was sufficiently concerned to do anything about it. Not only was the matter outside their core area of responsibility but also the community as a whole benefited from the trade, while the key organisational figure in this illicit timber traffic was none other than the desa head himself.

Seeing the remarkable success of the timber trade at Telaga Biru, some entrepreneurs at nearby Sepulu decided that they too should become involved and make their small town a conduit for the Java timber market. As at Telaga Biru, the key figure in this decision was the head of the Sepulu desa, in this case a man named Haji Komar. Haji Komar was a strong and charismatic person whose word was regarded as law. He had been desa head since the mid-1970s and throughout his long period in office no-one had dared to criticise him publicly, nor to stand against him in the periodic desa elections. His extraordinary authority and the respect with which he was regarded depended not on his wealth or education but on the force of his personality. He was an example of a strongman (*blater*) local leader, an exception to the pattern in Madura of desa heads who have little authority outside their immediate hamlets.

The personal style of leaders like Haji Komar is far removed from the typically gentle mien of the undisputed elite of Madurese society, the kiai. Nevertheless, tough formal leaders can sometimes gain widespread support through their ability to maintain law and order and to foster the local economy. Moreover, despite sharp outward differences of style, there is no conflict with the religious leadership. Because they are not expected to engage in business dealings, kiai are likely to depend on the economic leadership and patronage that a strong formal leader can provide, while the latter for his part depends upon the support of the local kiai for his continuing mandate (Muthmainnah, 1998: 27–8).

Haji Komar saw the opportunity and seized it. With the decline of Sepulu's supremacy in the fish wholesaling trade, the local economy had become stagnant and increasing numbers of people were leaving to try their luck in Kalimantan. Moreover, this was happening at a time when the economy in Java was surging ahead under the New Order's developmental program. A few vessels from Sepulu were already involved in timber trading between Central Kalimantan and Pasuruan. If that timber were to be discharged in Sepulu and sold to buyers from Java, the whole local community could benefit. The economic advantage over the Pasuruan market would not be from dispensing with SAKO documentation because the Sepulu vessels carrying timber to Pasuruan had never bothered to obtain a SAKO anyway. The advantage would be logistical as it was only one and a half days by motorised prahu from Sepulu to the coast of southern Kalimantan, while from Sepulu it was only about three hours by road and ferry to Surabaya, where the demand for timber was almost insatiable. With substantially lower operating costs by comparison with landing timber in Pasuruan, and no port costs or administrative problems to worry about — there was then no harbourmaster at Sepulu — the timber could be well priced to compete on the Surabaya market. Needless to say, there was no intention of bothering with

SAKO documentation. The state had not looked after Sepulu. As had always been the case on Madura, the people would have to look after themselves.

The decision to make Sepulu a timber port was a great success, with a stream of buyers from Java eager to buy timber at discount prices. There was no market competition with Telaga Biru. The traders of Telaga Biru offered high-quality species such as bangkirai and benuas, while the wood landed at Sepulu consisted mainly of cheaper varieties such as meranti and kapur which were in most demand for interior construction purposes. These cheaper species were also the main commercial species anyway in the areas where the Sepulu vessels obtained their timber, whereas bangkirai and benuas came from much further west.

Only Sepulu vessels were allowed to land timber at Sepulu; vessels from elsewhere could be beached there for maintenance but they could not discharge timber. The Sepulu entrepreneurs obtained their timber from the Katingan river, to the east of the Mentaya, and other smaller rivers flowing into the deep bay to the east of the Katingan (Map 13). This sparsely populated area had been little exploited during the early decades of the timber boom because it was very difficult for sailing craft to access. On both sides of the entrance to the Katingan there are shoal banks extending far offshore with much broken water during the southeast monsoon. It is very easy for a vessel to run aground here because, apart from being shallow, the water is so murky that the bottom is not visible even in just a few inches of depth, while the channel between the shoal banks is narrow with no navigation aids other than the odd crude pole. However, the main difficulty for engineless prahus was not the entrance but making way upriver against the current. Even with an engine it took up to eight hours motoring to reach the sawmill area, near Mendawai.

With most vessels opting instead to load timber from the more accessible Mentaya river or from Kumai further to the east, the Katingan-Mendawai area was reputed in the early 1990s to have the most plentiful remaining timber supplies of any river system in southern Kalimantan.[11] As this resource began to be exploited by Banjarese

[11] In 2003 some of the very large new Bugis vessels, carrying over 1,000 cubic metres of timber, were loading their cargoes mainly from the Katingan. The Katingan was also by then the major source area for logs which were carried by barges, towed by tugs, to Gresik for processing into plywood.

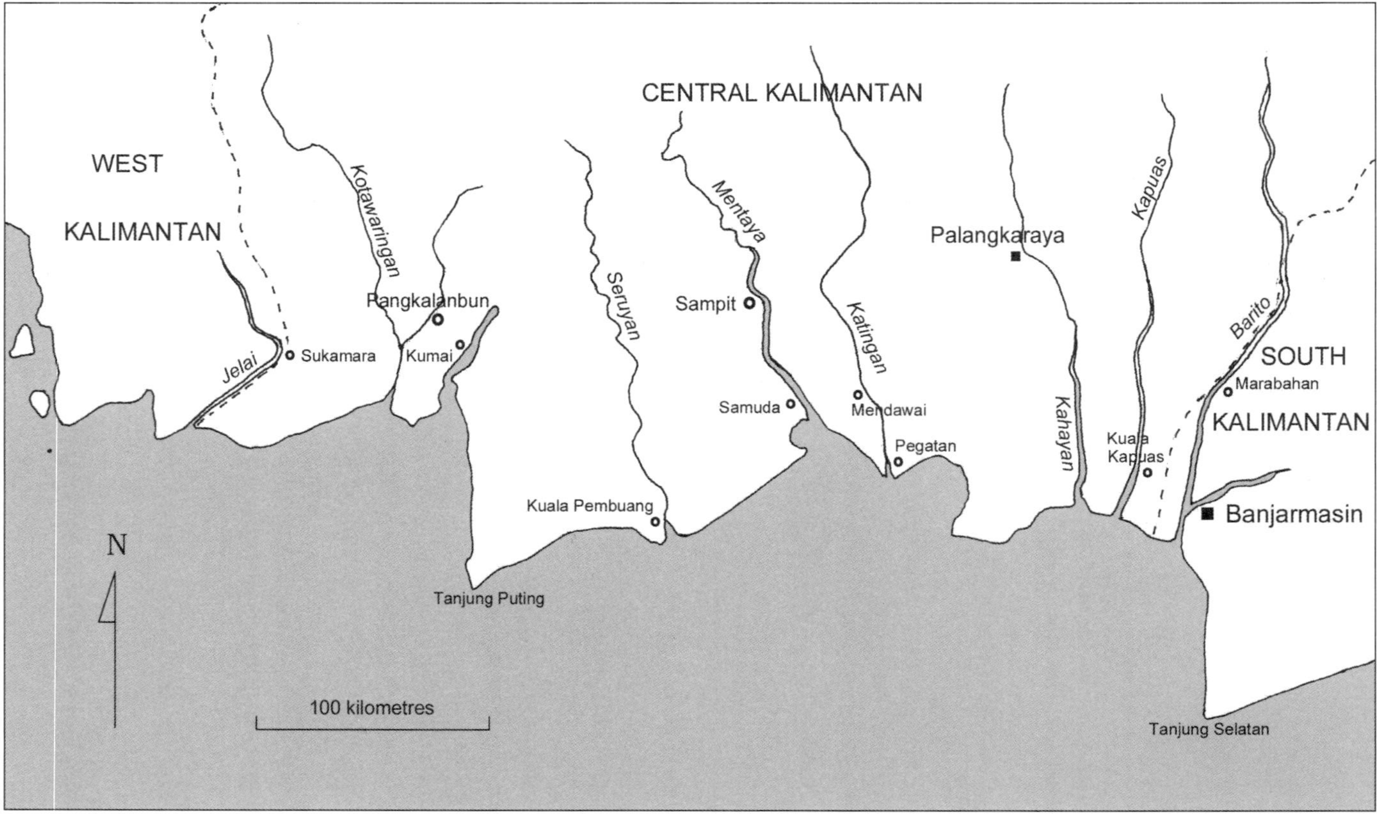

Map 13 The main waterways and ports of Central Kalimantan

sawmilling entrepreneurs, the Sepulu prahu operators took advantage of the opportunity. With rapid turnarounds at both ends and the proximity from the north coast of Madura to the south coast of Kalimantan, the Sepulu vessels could make four voyages a month, compared with only one voyage per month for most vessels discharging in Pasuruan. With so much prahu traffic in and out of Sepulu, in 1993 a harbourmaster's office (a branch of the Telaga Biru office) was established there.

As local vessel owners and timber traders prospered, new prahus began to be built at Sepulu. Unlike at Telaga Biru, where all vessels built after 1983 had been of 'modern' motorship styling, these new motor prahus at Sepulu were built after the *golekan* style with the characteristic scrolled ends and decorative paintwork. To maximise capacity, however, they were much beamier and deeper-bodied than in the past and had large box-like deckhouses. There was no sailing rig. Because teak was by then too expensive, supplies of *bungur* (*lagerstroemia* sp.) timber were purchased from Central Kalimantan.[12] Although an excellent marine timber, bungur could be purchased through special orders for not much more than the price of meranti. Employment opportunities boomed in building vessels, crewing, unloading timber on to the beach, loading it on to trucks, and in sawmills. Decrepit fishing boats were put into service to carry timber from prahu to shore, food stalls sprang up on the foreshore to serve the men working on the beach, and even local *becak* operators did much brisker business than before. With the beach and surrounding groyne covered in stacks of timber, bandsaws whining in makeshift outdoor sawmills in the residential area (to convert large pieces to smaller stock to meet the demand for local house building timber) and a steady procession of timber trucks, it was clear that Sepulu's economy now rested not upon fish but timber.

Although not a seafaring man himself, Haji Komar invested in vessels of his own and soon prospered. Yet although he was the key figure behind Sepulu's economic boom, Haji Komar's personal success lagged behind that of some other locals, especially his counterpart in the desa of Prancak, the neighbouring ward of Sepulu. Prancak adjoined the beach and in its outward character was much like a

---

[12] At Sepulu this timber is called *langoting*, a name borrowed from Dayak and Banjarese speakers. Bungur is the standard term in Indonesia for the same species of timber.

fishing village, whereas Sepulu was more obviously commercial with an open street market and shops. The head of the desa of Prancak, Nur Habib, was the wealthiest person in the town. Having made his fortune as a fish wholesaler, he now became the biggest operator in the timber business with five vessels to his name. With the leadership and inspirational example of these two desa heads, the economic progress of Sepulu seemed assured.

Following the success of Sepulu and Telaga Biru, two more conduits for the Java timber market emerged on Madura. One was the village of Batioh in the kecamatan of Banyuates, about 25 kilometres to the east of Telaga Biru. There was a narrow inlet at Batioh and a small fleet of *golekan*, by this time all motorised, was based there. The other was the village of Pagar Batu, on the south coast of Sumenep directly opposite the island of Giligenting (see Map 9, Chapter 3). Although this was well removed from the northwest coast, some entrepreneurs from the northwest were involved in financing the traffic of illegal timber through Pagar Batu.

With the national economy buoyant and the demand for construction timber steady, the mid-1990s was a heady time in these timber smuggling centres on Madura. It was not only prahu owners who prospered. Many of the wealthier people from the local and hinterland communities became involved as financial backers for timber voyages because the returns available from these ventures were well beyond what was obtainable from any other form of investment open to them. Yet throughout this period there was little awareness in these circles that this new prosperity rested on a fragile base. The halcyon period for the prahu operators and timber traders of Madura was in fact drawing to a close. From 1997 onwards a much less benign business climate would prevail. Four new factors would seriously affect the economy of Madurese prahu operators and timber traders: a sustained official crackdown on timber smuggling, the Indonesian monetary crisis that began in late 1997, the ethnic conflict in Central Kalimantan and the introduction in 2001 of regional autonomy.

The Crackdown on Timber Smuggling

Although the main prahu ports of Jakarta and Surabaya had been targeted over the years by the authorities to ensure some degree of compliance with the timber transport regulations, little concerted action had been taken elsewhere. Confiscation of cargoes had occurred from time to time but it was only in the second half of the 1990s that such decisive action began to be carried out in timber ports all along the north coast of Java. In February 1996 the Forestry Minister stated that 70 per cent of the timber coming into Jakarta was illegal. Subsequently 83 vessels were detained in the capital's prahu ports for carrying undocumented cargoes (EIA, 1998). In 1997 timber confiscations became widespread. The government had been coming under increasing pressure from environmental groups, both within Indonesia and internationally, to show that it could responsibly manage the forest reserves of Kalimantan. This pressure had been building up for years but it gained impetus in 1997 because of the region-wide smog caused by forest fires in Kalimantan and Sumatra. Air pollution from these fires blanketed not only western Indonesia but also Singapore and West Malaysia, embarrassing the Indonesian government within its own region. It was difficult to monitor uncontrolled burning in the forest areas so the government responded to the crisis instead by cracking down on illegal timber imports to Java. There was no direct connection between illegal timber shipments and the forest fires but a perception was nevertheless created that the government was taking a firm and responsible approach to managing the forest environment. Throughout the latter part of 1997 news items appeared frequently about the apprehension of timber smugglers (*penyelundup kayu*), as if such smuggling (evasion of taxes on imports) was a new phenomenon. All the main timber ports of Java were targeted in operations conducted by police and forestry task forces. In a joint operation in 1997 in the port of Juwana, for example, 38 prahus were reportedly found to have cargoes without proper documentation, with 10,500 cubic metres of timber being confiscated as a result.[13]

In another large-scale police operation, at Pasuruan in November 1997, 50 prahus were reportedly detained and had their cargoes seized.

---

[13] A shipping agent at Juwana claimed that only 4,500 cubic metres of this confiscated cargo was transferred to the Forestry Department to be auctioned, the remainder 'disappearing' to the benefit of corrupt officials. The same agent also purchased 1,000 cubic metres at the subsequent auction on behalf of a timber merchant in Semarang. He paid Rp 365,000 per cubic metre, a fair market price. However, he claimed that the storage and transport costs involved were greatly exaggerated by the officials involved, with the result that the net revenue for the government was a mere Rp 23,000 per cubic metre. Again, the difference allegedly went to corrupt police and forestry officials.

According to informants, a red 'X' was painted on the hull of each offending vessel and the vessels' passes were all seized by the police. These essential documents were apparently never returned, obliging the prahus to wait in port for two months with no income. Eventually the skippers had no choice but to depart without their vessel passes and to obtain new ones in their home areas. At a reported Rp 80,000 for a replacement document, the cost was not prohibitive but the inconvenience and time spent to obtain it was a considerable burden.

As a result of this official crackdown on revenue evasion, many prahu operators who for years had been using Pasuruan became afraid to land timber there. In January 1997 there had been 344 vessel calls at Pasuruan: in January 1998, after the confiscations, only 133 vessels entered the port.[14] However, although water police (*polisi airud*, water and air police) teams continued to check vessels arriving in Pasuruan for months afterwards, there were few confiscations after the initial onslaught of November 1997. This was possibly because the police teams involved had realised that further drastic action such as in the November raid could leave little opportunity to extract money in the form of *pungli* (*pungutan liar*, illegal levies). The preferred method of 'policing' now became the imposition of a so-called 'sea tax' (*pajak laut*), typically Rp 500,000 per vessel, in the Pasuruan roads before the vessel was allowed to enter the port. According to skippers, no receipts were issued for these unofficial charges. There were also many claims of partial confiscation of cargoes en route to Pasuruan, such as that suffered by the vessel owned by one Pak Roshid, of Sreseh, which was intercepted by a police speedboat off Puteran Island, near Kalianget, in January 1998. When it was found that there was no documentation on board for the cargo, the vessel was ordered into Kalianget, where eight cubic metres of timber (10 per cent of the total cargo) were unloaded under police supervision. After that, the vessel was allowed to proceed. According to the vessel owner, no receipt was given for the timber taken, nor did such a 'fine' provide any assurance of immunity from further impositions upon reaching Pasuruan. The loss of 10 per cent of the cargo would have left the voyage virtually profitless. Other skippers had similar stories.

[14] From records of port entries, Pasuruan harbourmaster's office. The decline in vessel calls at Pasuruan probably also reflected a drop in consumer confidence after the crash of the rupiah in late 1997.

This sustained crackdown by the authorities caused consternation among vessel owners, their financial backers and crews. This was especially so in Sreseh because Pasuruan had been the discharge port for all Sreseh prahus. There seemed to be genuine confusion in Sreseh as to why Pasuruan had been targeted. Most put it down to official corruption. No one conceded that the cargoes were illegal. Everyone knew that timber was supposed to be documented but in practice this requirement had been so little policed in the past that it had seemed irrelevant. Moreover, no one in Sreseh seemed to know exactly how the documentation process was meant to work. Some skippers said that the fees required to obtain the SAKO in Kalimantan were excessive, while claiming that even if they did pay them, the resulting documentation would still not be accepted in Java. This was probably incorrect but the skippers may also have been misled by officials who were more concerned to get money from bribes to 'let the shipment go' than to educate the skippers about the proper procedures. However, as will be shown, to have complied fully with the timber transport regulations would have resulted in significant disadvantage on the highly competitive Pasururan timber market.

The impact of the police campaign in late 1997 would in any case have been severe but it became devastating because these confiscations took place just as the rupiah was beginning its plunge against other currencies — the onset of the prolonged Asian monetary crisis (*krismon*). A typical case of an entrepreneur who was profoundly affected by this combination of events was Haji Ansuri of Batuputih, Sreseh. This man, a highly respected person in his community, had borrowed gold to the value of ten million rupiah, with which he planned to buy a cargo of timber from Pegatan-Mendawai (on the Katingan). His financial backers were not people from his own small village of Batuputih but rather friends and relatives from Sampang and Bangkalan, non-seafaring people, who lent gold jewellery and heirlooms. Unfortunately for the venture, when Haji Ansuri's prahu arrived in Pasuruan the entire cargo was confiscated. Consequently, he had no proceeds with which to repay his debt. Worse, because his backing had come from gold, it had to be repaid to the current value of that gold. As the rupiah plummeted in value against gold, his debt blew out to Rp 40 million, a huge amount in the local economy. A few individuals in the nearby village of Taman, where there were several large prahus with a capacity of over 200 cubic metres of timber, had incurred even larger debts. One Taman vessel owner was said to owe Rp 180 million. In this society there is no option of bankruptcy: a debt remains as long as it takes to discharge it. Some entrepreneurs went off to work in Saudi Arabia, the only route by which they could extricate themselves from such massive debts. They could not sell their prahus because with no-one prepared to invest in the timber business, there was no longer a demand for such large vessels. Many of these big prahus were simply left derelict on the beach.

Apart from the serious financial risks involved under the new timber policing regime, there were other factors clouding the future of small-scale maritime entrepreneurship. Consumer confidence was weak as a result of the economic crisis, which clearly was not going to be overcome in the short term, leading to a slowdown in the construction industry and reduced demand for timber. Moreover, as a result of the sharp decline in the value of the rupiah, engines and all associated equipment were now much more expensive because they had to be imported. Most large prahus by this time had engines, now regarded as almost essential to reach ports such as Pegatan, but the cost of fitting out a new vessel now rose dramatically. In the case of Sreseh, where building of new prahus had continued steadily throughout the mid-1990s, construction came to a standstill in 1998 except for the completion of two vessels that had been started in the previous year.

Other than short-term contract work overseas, which in any case was difficult to obtain and by no means available to everyone, there were no acceptable employment alternatives in Sreseh. The salt trade, the former economic mainstay for the local prahu owners, by this time presented only limited opportunities. Prahu owners could still undertake timber transport work, provided they could find a trader in Pasuruan willing to employ them as carriers in this time of reduced demand for timber, but the income from freight work alone was much less than if one also traded on one's own behalf. Haji Ansuri lamented, "We are a seafaring people. How else can we live, except by carrying timber?" However, it was clear from the course of the conversation that what this unlucky maritime entrepreneur really meant was trading in timber as well as carrying it.

The requirement that SAKO documentation be obtained for all timber landed should not have made any significant difference to the nature of the timber trade. The additional cost of obtaining a SAKO could, and should, have been borne by consumers. All that was needed was for the timber traders to stand together and act in their collective interest, so that none would attempt to undercut the others by selling cheaper timber for which no SAKO had been obtained. However, in the busy port of Pasuruan, dominated by Madurese entrepreneurs, this possibility was never suggested. The Madurese are well known historically for their lack of communal action and this case was no exception (Kuntowijoyo, 1980: 526). Although the flood of timber without any documentation had been halted, there was still plenty of illegal timber being landed at Pasuruan, in most cases with extra-legal payments in cash or timber to the water police along the way.

In Sepulu, however, the Madurese pattern of individual action with no concern for the collective good did not prevail. The authorities had finally become aware of the volume of timber arriving illegally in Madura and Sepulu, as the main timber gateway on the island, now became the primary target for police and forestry teams. Nevertheless, Haji Komar, the driving force behind the phenomenal economic growth of Sepulu, did not despair like the entrepreneurs of Sreseh. Realising that it was vital for the Sepulu timber merchants to stand together, in late 1997 he founded an association, the Persatuan Pengusaha Sepulu (Sepulu Entrepreneurs' Association). Office premises were made available for this new association and an administrative officer appointed. Membership was compulsory for all vessel owners based at Sepulu, with a basic monthly levy as well as a charge for each truckload of timber departing the port. At first some were unhappy about paying this money, even though they knew it was in their collective interest, but no one dared defy Haji Komar. In 1997 the charge per truck was Rp 75,000. With the funds collected, pay-offs were organised, both on a retainer basis and for each shipment by truck, to a web of police and forestry officials. This system of organised graft was necessary to keep Sepulu free from the same sort of sustained law enforcement offensive that was being conducted in Pasuruan and elsewhere, but it was possible only because of Haji Komar's charisma and authority.

In general, 1998 was a grim year for timber transporters and traders operating on the margins of legality. Yet gradually a pattern of partial compliance with the regulations emerged, with most entrepreneurs opting to pay for a SAKO of between 30 and 40 per cent of the actual amount of timber carried. In this there was an element of bluff, as few people not involved in the trade could reliably estimate the amount of timber carried by a prahu. The measurement certificate (*surat ukur*) is at best a vague guide in this respect. Many small prahus sailing to East Java did not carry a measurement certificate anyway, with agents relying instead on the vessel's pass (*pas kecil*), often showing a greatly understated gross tonnage (see Appendix 3). In any case, partial compliance of 30 to 40 per cent was enough to ward off the financial catastrophe of complete confiscation of a cargo. Lesser charges, such as the illegal 'sea tax' routinely exacted by water police patrols from prahus calling to Pasuruan, could be borne. The 'sea tax' was a substantial imposition and prahu operators were bitter over it. Nevertheless, they knew, as did the police patrols, that it was much cheaper than the alternative of paying for a SAKO for the entire cargo.

Although the rupiah remained extremely weak, in 1999 consumer confidence began to return. In the second half of the year the timber trade boomed again, with a record 405 timber-carrying vessels arriving in the port of Pasuruan during the month of October.[15] Despite frequent visits to the port by water police patrols based at Surabaya, the level of compliance with the timber transport regulations remained low and many entrepreneurs resumed financing shipments without any documentation. This ignored the passing during the year of a new Basic Forestry Law (Law 41/1999), which provided for much stiffer penalties for illegal logging — up to ten years in prison and fines of several billion rupiah. Under the new forestry law, the SAKO was replaced by the SKSHH (certificate of legality of forest produce) but this administrative expedient made no difference to the business operations of timber shippers.

The following year saw the authorities take a tougher stance, involving more confiscations of illegal timber cargoes. In Taman, Sreseh, several owners of large prahus experienced calamities of this nature. Among these unfortunate entrepreneurs was one Haji Latif, whose *janggolan* was caught in Semarang in 2000 with no SKSHH, resulting in confiscation of the entire cargo and a personal debt of around Rp 60 million. In another case from Sreseh during the same year, a prahu that had been launched just a few months earlier was intercepted off the western end of Madura by a water police patrol. The prahu was carrying 50 cubic metres of timber from Kumai with no SKSHH. According to crew members, this was standard practice for vessels loading in Kumai at that time. With the owner-skipper, Haji Abdullah, of Batuputih, unable to provide the necessary documentation, the prahu was ordered into Surabaya and directed to

---

[15] From records of port entries, Pasuruan harbourmaster's office.

berth in an area controlled by water police. There the vessel and crew were detained for a week. According to a crew member, the police were abusive towards them, they were not allowed to use an outdoor bathroom/toilet on the wharf, and threats were even made to the skipper that the women in his household would be molested. Eventually the vessel was allowed to retain its cargo after the merchant who had financed the purchase of the timber arrived from Pasuruan and paid a reported Rp 5 million to the police.

With incidents such as this becoming commonplace, few people in Sreseh wanted to invest in new vessels. On the other hand, in Sepulu, where the flow of timber continued unabated with seeming immunity from official interference, the demand for vessels remained high. New vessels were still being built there, while some Sepulu entrepreneurs bought existing vessels from elsewhere. One who did so was Haji Komar himself, who purchased two nearly new prahus in 2000 from Sreseh. Both of the vendors made handsome profits on their deals and both made the pilgrimage to Mecca shortly thereafter.[16]

For an owner to sell a newly completed vessel for a handsome profit is unusual. The Madurese are renowned for their frugality but in Sepulu, the economic miracle of the island, attitudes towards consumption were changing. The most conspicuous sign of such change was the building of several grand houses in the small town. Some of these large and lavishly appointed residences were said to cost over one billion rupiah (about $US 100,000), a huge amount by the standards of ordinary people. Other less obvious signs of social change as a result of the booming local economy included small cheap restaurants selling beer (a proscribed beverage in West Madura) as well as food, a billiard saloon, and even the odd transvestite appearing in public. Further, earnings from manual and casual labour were significantly higher than elsewhere on Madura, and indeed probably higher than anywhere else in East or Central Java. The few *becak* operators did a brisk trade, earning much more than their fellows in the much larger towns of Sampang and Bangkalan, while skilled boatbuilders were commanding a daily wage of Rp 70,000, compared with Rp 30,000 in Sreseh.

---

[16] At that time the cost of the pilgrimage from Indonesia was around Rp 25 million per person (about $US2,500), a large amount of money by local standards. A typical unskilled manual worker in Java was then receiving only about Rp 10,000 ($US1) per day.

What enabled this rising standard of living was of course the continuing flow of timber through the beach port, most of it still illegal. The presence of a harbourmaster's office in Sepulu was no impediment to the illicit traffic. The harbourmaster was aware that timber transport regulations were not being complied with and claimed that he routinely noted the absence of timber documentation in his reports to his superiors in order to protect himself, but no further action was taken. As he put it, he had no choice but "to go along with the culture".

The scale of operations at Sepulu was described to the author in 2001 by Hamid Sahal, an enterprising young Madurese timber trader from Pasuruan. Hamid's business was not large, being supplied by only one prahu, and from time to time he received orders beyond the capacity of his stock. On such occasions, he would obtain the necessary extra timber from Sepulu. According to Hamid, approximately 100 trucks would leave Sepulu every night, each truck carrying 10 to 16 cubic metres of timber. Sometimes there were many more trucks than this, he claimed. In addition to the cost of the timber, he had to pay the standard 'insurance money' of Rp 85,000 per truck, which went into the fund administered by the Sepulu Business Entrepreneurs' Association. From this fund, he claimed, regular payments were distributed to a wide range of officials: water police, navy personnel, forestry officials and, most importantly, ordinary police, with allocations for the different police administrative tiers of *sektor* (kecamatan or sub-district level), *resort* (kabupaten level) and *wilayah* (Madura-wide level).

The traffic from Sepulu of around 100 trucks per night was subsequently confirmed by fieldwork observations, as was the amount paid per truck as 'insurance'. All entrepreneurs paid the same amount (by 2002 it had risen to Rp 95,000). The Association was thus receiving in the vicinity of Rp 250 million per month throughout the sailing season, in addition to the standard monthly contribution paid by local entrepreneurs, with nearly all of this money being used to ensure the unhampered flow of timber into Sepulu and from there into Java. This facilitative process was more than a matter of just bribing officials not to pay attention to the legal status of the timber: each truckload would be provided with documentation to show that its load was fully legal, regardless of its prior history — a process known as 'whitening' (*pemutihan*). This complex system of pay-offs organised through the Association worked with such efficiency that the entrepreneurs of Sepulu were the envy of those vessel operators and traders who had to regularly run the gauntlet of police, navy and forestry officials at Pasuruan. In 2001, however, a serious problem of a very different nature flared up. This time the strategy of graft payments would be quite inadequate as a solution.

The Ethnic Conflict in Central Kalimantan

Unlike in West Kalimantan, where there had been a long series of violent incidents between Madurese and Dayaks leading up to the conflicts in 1997 and 1999, in Central Kalimantan there had been no history of serious violence between the two groups before 2001. Anti-Madurese sentiment became widespread only in 2000, after a bitterly contested election in the provincial assembly for the position of governor. Since the early period of the New Order there had been resentment among Dayak elites in Central Kalimantan over the appointment of Javanese figures to the provincial governorship. After the collapse of the New Order, an opportunity for a Dayak to become governor came in 2000. The winner of the gubernatorial election of that year, Asmawi, was a Dayak, as was the candidate who came second, Professor K.M.A. Usop, the leader of a group with the specific aim of achieving Dayak control of the province. Having lost the election, Usop and his group thereafter did its best to discredit Asmawi and to undermine his position as governor. Failing to dislodge him, Usop's group then seized the chance for a different approach to power. After a Dayak man had been killed by three Madurese in a small mining town on 15 December 2000, Usop and other leading members of his group visited the area concerned and turned the incident into a symbol of the oppressed condition of Dayaks in their own country, warning Jakarta that Dayak anger towards the Madurese could not be contained much longer. This was, in effect, the launching of a campaign for a 'cleansing' of all Madurese from Central Kalimantan, making the Madurese the scapegoats for all the problems faced by ordinary Dayaks. Just two months later, on 18 February 2001, a group of about 50 Dayaks attacked Madurese in Sampit, killing 6. The Madurese fought back and took control of the town for two days, before being overwhelmed as truckload after truckload of Dayak fighters were brought in, probably from mining and logging areas. Hundreds of Madurese were murdered, many of them beheaded, culminating in the exodus of over 100,000 refugees to Madura.

Meanwhile, Usop was being hailed as the new champion of the Dayak people (van Klinken, 2002; *Forum Keadilan*, 49, 18 March 2001: 19).

No convincing evidence has been presented of intense anti-Madurese sentiment in Sampit.[17] However, the initial attack came just one day before a planned bureaucratic reorganisation in the kabupaten of Kotawaringin Timur, of which Sampit was the administrative centre. It was reported that some new appointees to the bureaucracy were not aligned with Usop's Dayak supremacist group, while the Sampit secretary of that group, a person who had played a coordinating role in the Dayak attacks, had lost his position in the administrative downsizing (*Tempo*, 11 March 2001: 23).

The Madurese were the obvious choice as scapegoats. First, as a migrant group they were very conspicuous. While they comprised only 6 to 7 per cent of the total population of Central Kalimantan, in Sampit, the largest town in the province, about 70 per cent of the population of around 120,000 were Madurese. Second, many Madurese had prospered in Central Kalimantan, arousing economic jealousy among Dayaks. Third, even though many had migrated to Kalimantan of their own accord, outside the government's transmigration program, they were inevitably associated with the New Order regime. Fourth, the history of ethnic clashes in West Kalimantan involving the Madurese and their expulsion from that province supported the notion, spread by Usop's group, that the Madurese were aggressive and unsympathetic towards Dayak culture.

Whereas the ethnic conflicts in West Kalimantan in 1997 and 1999 did not significantly affect Madurese involvement in the timber trade,[18] the 2001 conflict had serious repercussions for many Madurese prahu operators and timber traders. For two decades Sampit had been the most important timber outlet in Indonesia. Many Madurese

---

[17] Refugees from Sampit told me that Madurese-Dayak relations in Sampit had been good. In Sampit, one year after the conflict, locals were guarded in talking about the violence. However, some said that the town used to be better when the Madurese were there because the economy had been stronger.

[18] The West Kalimantan conflicts in 1997 and 1999 did, however, disrupt the cattle export trade from Telaga Biru. Hitherto the cattle slaughtering and butchering trades in West Kalimantan had been dominated by migrant Madurese (Sudagung, 2001: 146). After the relocation of most of the Madurese population of the province to refugee centres, the cattle exporters of Telaga Biru were forced to diversify and seek alternative markets in Bangka and Beliton. The cattle trade to Pontianak subsequently recovered.

in Central Kalimantan had become involved in the timber trade and Madurese prahus had been a common sight on the Mentaya river. Now, in the aftermath of the orgy of slaughter in Sampit, no port along the Mentaya could be considered safe for Madurese vessels. Even Samuda, almost at the entrance to the Mentaya, was dangerous. Although Samuda had remained peaceful while fighting and killing raged in Sampit, Dayak mobs had eventually arrived in trucks and attacked the Madurese residents there (Surata and Andrianto, 2001: 110). The Katingan, Sebangau, Kahayan and Kapuas rivers were similarly unsafe. Relations between Madurese and Dayak in timber loading areas along these waterways generally seem to have been good but there were many stories of Madurese in rural areas being hunted down and killed by Dayak groups from outside the areas concerned. Fear of such murderous roving gangs was enough to keep Madurese vessels away.

This had a severe impact on the economy of prahu operators from Sepulu and Sreseh, who prior to the conflict had been obtaining their timber mainly from the Mentaya and Katingan rivers. Prahu operators from Giligenting and Sapudi were much less affected because many of them obtained their cargoes from Kumai and Sukamara in the western part of the province. This area was only lightly touched by anti-Madurese violence, perhaps because there the influence of Usop's extremist group was not so strong.

Besides the safety issue, after the Madurese exodus most Madurese timber traders in Pasuruan lost their supply contacts in Central Kalimantan, because in most cases the sawmillers they had relied on were also Madurese. Eventually most of the Madurese vessels discharging at Pasuruan and which, before the conflict, had been obtaining their timber in Central Kalimantan, turned to sources in South Kalimantan. Although this transition was not difficult, the timber was more expensive than in Central Kalimantan and vessels had to wait longer to obtain their cargoes because supplies were less plentiful.

This turn of events was particularly worrying for the entrepreneurs of Sepulu, whose profits had depended on cheap timber and up to four voyages per month. After several months during which the local timber trade remained at a virtual standstill, some Sepulu vessels did try to load at Pegatan, on the Katingan. They were well received there, but it was not long before an attack was carried out by a Dayak gang from outside the area, murdering an innocent Madurese crewman.[19] In other cases murderous intent was replaced by common banditry, with Madurese crews becoming victims of extortion by Dayak gangs demanding large sums of money to ensure their safety.

Faced with this very difficult situation, the entrepreneurs of Sepulu came up with a remarkable strategy: to conceal the ethnic identity of their vessels. There were, after all, many Bugis vessels obtaining timber from the Pegatan-Mendawai area. First, it was decided that no more *golekan*-style prahus should go to Central Kalimantan. Only these vessels, with their unmistakeable Madurese styling, had been targeted by Dayak bandits. All *golekan* instead now turned to South Kalimantan for their cargoes. Furthermore, although there was strong loyalty to the type in Sepulu, no new vessels of this type were built thereafter. Vessels of the *sepel* model, some of which had been purchased from Sreseh and which closely resembled smaller Bugis and especially Mandar craft, were much less likely to attract unwelcome attention. There were also a few vessels at Sepulu styled after the small motor-ships of the Riau port of Bagansiapiapi (Lips, 1993: 92–102), which were likewise not generally associated with the Madurese. Second, in order to replace the *golekan* vessels lost to the Central Kalimantan trade, vessels were chartered from other places, especially Bima, in Sumbawa. Such arrangements were made in Gresik and Pasuruan, where Bugis vessels from Bima were frequent callers. Third, many non-Madurese sailors, mainly Bugis, were hired in Gresik and Surabaya to replace Madurese crews. In most cases, however, vessels carrying timber from Central Kalimantan to Sepulu thereafter did have at least one Madurese crew member, whose ethnicity was kept secret when in Kalimantan. This 'trusty' was rarely the skipper, as this would have been too dangerous. Most Madurese prahu sailors have only an elementary education and their spoken Indonesian is typically rough with a heavy accent that could easily have betrayed their ethnicity to a sympathiser with Dayak extremists. Accordingly, the skipper, responsible for all negotiations on shore, was in most cases Bugis. The Sepulu sailors so replaced were not necessarily unemployed, most obtaining positions on *golekan* sailing to South Kalimantan.

[19] I had previously heard about this incident from sailors at Sepulu but it was confirmed by local officials at Pegatan Hilir, in Katingan Kuala. Apparently the perpetrators had come from upriver and were not from Katingan Kuala. The attack was ethnically motivated with no connection to the timber trade.

Photo 30 Unloading timber on the beach at Sepulu, 2002. The vessel in the foreground has been chartered from Bima.

As a result of this strategy, Sepulu's timber business continued to thrive. New vessels were ordered, especially from Sreseh, where labour was cheaper than in Sepulu. In September 2001 several vessels were under construction in the tidal creek at Bupoteh, Sreseh, for Sepulu owners. In one case the work gang had been contracted from Sapudi. For an owner in Sepulu to bring a gang from distant Sapudi to build a vessel in Sreseh, on the south coast, seems on the face of it an unlikely arrangement, but with the beach at Sepulu almost covered with stacks of timber, little space was available there for the building of large vessels. While Sreseh was not exactly close to Sepulu, it was at least possible for the owner to be able to visit from time to time to check progress and keep an eye on the construction.

During the year after the ethnic conflict, the situation in Central Kalimantan gradually became safer for Madurese prahus, with local police stopping the gangs of Dayak bandits who had been trying to extort money from Madurese crews. Large vessels from Sapudi and Giligenting were again regular callers to the Mentaya, especially at Samuda, and even *golekan* could again load timber in the Katingan without any problem. Nevertheless, with the exception of vessels from Sepulu, few of the smaller Madurese craft which had turned to South

Kalimantan for their cargoes returned to Central Kalimantan, because in the meantime a new and more intractable problem had emerged: the imposition of new regional taxes.

The Impact of Regional Autonomy

In 1998 and 1999 the Habibie government introduced a series of reforms aimed at decentralising government and giving greater autonomy to regional areas. The two most important of these reforms were Law 22 of 1999 on Regional Governance and Law 25 of 1999 on Fiscal Balancing, which together provided the main legal basis for regional autonomy (Barr and Resosudarmo, 2001). Under these laws, which came into effect in January 2001, many administrative and regulatory functions previously carried out by the central government were transferred to regional governments, with the key functional level being not the province but the administrative district or kabupaten. At the same time, each administrative district was expected to take steps to generate its own funds through the management of local resources instead of relying on assistance from Jakarta and the provincial government as in the past. To assist in the transition to this new system, in 1995 one district had been selected from each province to participate in a nation-wide pilot project for regional autonomy. In Central Kalimantan, the pilot district was the kabupaten of Kotawaringin Timur (East Kotawaringan), with its capital in Sampit. During the trial period from 1996 to 2000, the Kotawaringin Timur government had little success in increasing district revenue and there was serious concern about the future. Consequently, the members of the district assembly in Sampit decided, in a liberal interpretation of the Fiscal Balancing Law, to create revenue for their district in a much more assertive and innovative way than had been envisaged by Jakarta.

In this vast and sparsely populated administrative district, the forestry sector offered by far the best potential for increasing revenue. It was common knowledge that illegal logging was being carried out on a large scale in Kotawaringin Timur, with no revenue being garnered from this industry. A task force formed in March 2000 to investigate the extent of the illegal timber business discovered, in a period of just a few weeks, no fewer than 178 vessels with illegal cargoes of timber in the Mentaya river alone. Undoubtedly the number would have been much greater if the Katingan and Seruyan river systems, both in Kotawaringin Timur, had been similarly checked.

Although by law these illegal cargoes should have been confiscated and those responsible prosecuted, it was decided in the district assembly that the vessels would be allowed to proceed unhindered, provided that they paid a district timber levy calculated on a per cubic metre basis. On payment of this levy, referred to as a *retribusi*, a certificate would be issued giving the cargo legal status so far as the Kotawaringin Timur government was concerned (Casson, 2001: 10).

This move generated massive additional income for Kotawaringin Timur, with district revenue increasing by Rp 11.8 billion (approximately $US 1.2 million) in just the first three months of application of the new tax. Extrapolating this figure for the remainder of 2000,[20] it appears that total revenue for the district for that year would have been around Rp 62 billion, a 1,500 per cent increase over the annual average for the period 1995 to 1999, with more than half of this from taxes on cargoes that Jakarta regarded as illegal (Casson, 2001: 16). Undoubtedly, this huge increase in revenue from timber shipments was a factor in the bitter struggle for the provincial governorship.

Not surprisingly, this bold initiative from the Kotawaringin Timur government attracted strong criticism from Jakarta on the ground that it was legitimising illegal logging. There would of course have been no objection from the central government to a new district timber tax provided that it applied only to timber obtained within the pre-existing forest concession system, with any timber obtained outside that system being liable to confiscation and those responsible prosecuted. However, compliance with this expectation would have greatly reduced income for Kotawaringin Timur. After decades of what was regarded locally as exploitation by Jakarta, the district assembly was in no mood to compromise. The legal timber companies operating in Kotawaringin Timur were also upset by the new district timber levy, which they had to pay in addition to their royalty and reforestation fees. Moreover, they had to pay the *retribusi* at a higher rate than the illegal loggers were being charged. The illegal operators were also disgruntled because they now had to pay a considerable amount for

[20] After an initial openness from Kotawaringin Timur officials about revenue from this source, it was subsequently made clear that further information would not be made available without special permission from the *bupati* (Casson, 2001: 13). Other significant amounts of revenue for the district came from logging concessions, Inhutani III (a state-owned timber venture) and mining.

which they had not previously been liable, and for a document of dubious legal status in Java.

Such cargoes obtained outside the legal system but bearing the approval of the Kotawaringin Timur government were in fact not generally acceptable in Java. A team from Kotawaringin Timur visited Java in July 2000 to promote the legal basis for its actions (Casson, 2001: 12–13). Despite some concessions — for example, from the kabupaten of Blora, in Central Java, an important teak forestry area — it increasingly became the case that all shipments of timber arriving in Java required a valid transport permit (SKSHH). In law this had been necessary all along but with increased official vigilance and concern over revenue evasion as well as the state of the environment, by mid-2001 it had become a practical necessity as well. Apart from vessels discharging in Sepulu and to a lesser extent in Pasuruan, few now dared to carry timber without an SKSHH. Although this document was supposed to be issued only for legally obtained timber, after decentralisation and the huge increase in the scale of the illegal timber trade since the mid-1990s, the origin of the timber became less important than the payment of taxes. There seemed to be little that Jakarta could do except to ensure that the royalty (PSDB) and reforestation (DR) levies were collected, regardless of the provenance of the timber.

Kotawaringin Timur's extraordinary success in increasing revenue through the establishment of its own timber regulatory regime led to some other administrative districts in Central Kalimantan adopting a similar policy. Notably, this included Kotawaringan Barat (West Kotawaringan) with its timber outlet of Kumai. Nevertheless, significant differences remained across areas in the efficiency of collection of revenue from timber. In the case of Kotawaringin Timur, checking the documentation of cargoes leaving the Mentaya presented no great difficulty, with police inflatables darting out to intercept any passing timber carriers. In more remote places such as Katingan Kuala (Pegatan), however, there was more laxity in implementing the district timber tax law.

This situation changed during the second half of 2002, with the creation of new administrative districts in Central Kalimantan. In the case of Kotawaringin Timur, the original district was divided into three: a 'mother district' of Kotawaringin Timur with its administrative centre in Sampit, and two assistant districts (*wilayah pembantu*) — Seruyan with its administrative centre at Pembuanghulu, and

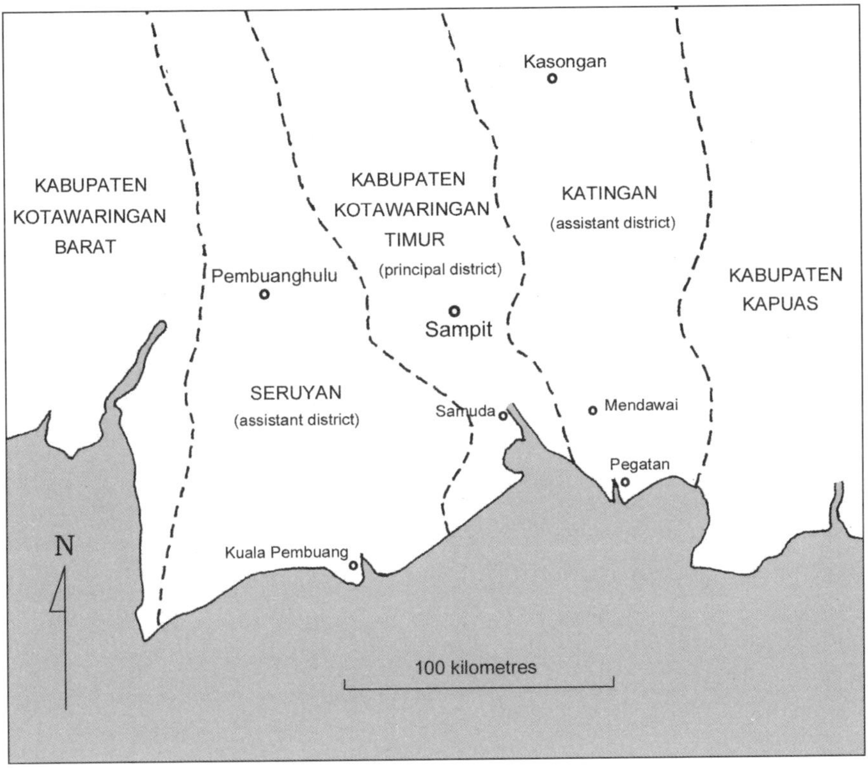

Map 14 Administrative reorganisation in Kotawaringan Timur, showing the new administrative districts excised from but affiliated to the principal district

Katingan with its administrative centre at Kasongan (Map 14). This move had been planned for some time to increase administrative efficiency in the outlying areas and to split revenues so that money garnered in one district would be to the benefit of that same district. Under the new arrangement, the salaries of administrative staff in the two new districts would be paid for only one year by the central government, after which time they would have to be paid from local revenue. Accordingly, the collection of taxes from departing timber vessels suddenly became a matter of vital importance for local officials. To assist in this matter, 'combined posts' (*pos terpadu*), consisting of representatives from the police, public prosecutor's office, district military command, forestry service and port officials were set up at strategic places such as Pegatan and Kualapembuang to check departing vessels. As a result of this new vigilance, it became much more difficult to evade payment of the district timber levy. At Rp 50,000 per cubic metre for sawn timber,[21] this levy was a significant imposition on purchasers and made timber from Central Kalimantan more expensive than timber from South Kalimantan. Consequently, most Madurese entrepreneurs bringing timber into Pasuruan preferred to obtain their supplies from South Kalimantan rather than Central Kalimantan.

Enforcement of the SKSHH

Although the increased cost of timber from Central Kalimantan was particularly vexing for Sepulu traders, by the latter part of 2002 this was not their major concern. More serious was an increased drive on the part of the authorities to curb illegal timber imports. From numerous accounts within the trade at both Sepulu and Pasuruan, this new strict policy began to be applied in both places during the latter part of 2001. Hitherto the police had been apparently unconcerned over partial documentation of timber cargoes, with 30 or 40 per cent being typical. Now they were insisting that SKSHH documentation should cover at least 90 per cent of the cargo — allowing a small amount for trading on the crew's own behalf — and accusing prahu operators of being 'recalcitrant' because of their reluctance to comply with the law.[22] In practice, vessels carrying an SKSHH for half or less of the cargo were still usually allowed through after making a substantial payment to water police, but the days when a vessel could slip into port with no documentation at all for its cargo of timber were over.

Checks on the legality of timber cargoes did not take place only at the destination port. A very fruitful place for police sea patrols was around Kalianget as most timber carriers bound for Pasuruan passed close to the eastern end of Madura. Many were caught there by police 'speedboats'. One case in 2001 concerned a sailing prahu from Giligenting, intercepted carrying 100 cubic metres of timber from Kumai with no documentation at all. The vessel was escorted to Kalianget and detained there, while the crew and skipper were allowed to go home to Giligenting. A week later, the skipper, a man of very modest means, was arrested by police and charged with transporting

---

[21] Personal communication, Drs Yusuf Sulaiman, Head of the Division of Planning and Development, Kotawaringin Timur, June 2003.

[22] Interview with staff at the *Airud* police office, Tanjung Perak, Surabaya, September 2002.

timber without a permit. The entrepreneur who had paid for the shipment, a man who lived at Banyuates on the north coast of Sampang, was also arrested and brought to Sumenep. Both the skipper and his timber 'boss' were sentenced to three months imprisonment. As the organiser of this timber smuggling venture, the boss also received a fine of Rp 38 million. Taking everything into account, however, his actual loss would have been in the vicinity of Rp 100 million.

As word spread of such incidents, most timber purchasers opted to pay for an SKSHH, albeit for much less than the full amount of timber purchased. However, during this early stage of 'socialisation' of the timber transport regulations in Pasuruan, some skippers still did not appreciate the necessity of carrying the SKSHH on board. The process of obtaining an SKSHH from the district forestry office usually took at least four days, notwithstanding that the cargo was rarely physically checked at the loading port. Since the SKSHH could not be applied for until loading had been completed, these four days were considered by skippers and crews as 'dead time'. Reluctant to wait around for so long, some impatient skippers naively arranged for the SKSHH to be sent by post to the consignee in Java, unaware that this was not acceptable to the authorities.

On a single day in June 2002, three vessels from Sreseh, each carrying timber bound for Pasuruan, were intercepted by a police sea patrol off Kalianget. None carried an SKSHH. In two of the three cases the permit had in fact been arranged and paid for (at 40 per cent of the actual volumes carried). In the case of the third vessel, however, no attempt whatsoever had been made to comply with the law. The owner of this third vessel, who was also the financer of the timber purchase, was one Haji Taufik, of Labuhan, Sreseh. This individual was widely regarded by many in the local community as excessively commercially driven, while from the viewpoint of the police he was a 'hardcore' offender, having twice before been caught for having organised undocumented shipments. All three skippers were arrested and placed in jail, along with Haji Taufik. The other two owners were not charged, although the SKSHH documentation was in each case for only 40 per cent of the total carried. Haji Taufik was subsequently released on a reported Rp 15 million bail but the three skippers, probably unable to raise any significant amount for bail, remained in jail. Due to problems with attendance of witnesses in court, the case against the skippers dragged on for three months before they were each found guilty of contravening the timber transport regulations and fined Rp 300,000. The result of the case against Haji Taufik is not known. However, in all three cases the timber cargoes, which had been unloaded and trucked away by the police as 'evidence' without any receipts being issued, had disappeared. One of the 'honest' owners, Haji Matsahri, claimed that as a result of this incident he lost a total of Rp 57 million. Haji Taufik's loss would certainly have been greater than this because his vessel was considerably larger and carried more timber.

With the penalty from such corrupt action being so far in excess of that imposed by the court, all Sreseh skippers thereafter took care to wait for the SKSHH to be received before departing port. Even so, they and many others discharging at Pasuruan remained vulnerable to police extortion because their documentation was typically for only 40 per cent of the total cargo. Moreover, many skippers and timber traders claimed that the demands of the police for 'sea tax' were becoming heavier. One timber trader spoken to in September 2002 said that he had just paid a bribe of Rp 2 million to police so that a shipment of his could be unloaded, even though an SKSHH was carried for 70 per cent of the cargo — by Pasuruan standards, an unusually high degree of compliance. Most traders felt that even if the SKSHH was for 100 per cent of the cargo, the vessel would still be subject to police extortion because hardly any of the vessels concerned complied with the Pelayaran Rakyat safety regulations. Most, for example, did not carry radios, notwithstanding that this was a requirement for all vessels with a capacity of 100 cubic metres and above. Consequently they remained easy targets for predatory corruption, regardless of the legal status of the cargo.[23]

As well as police, naval personnel were also apparently becoming involved in such predatory activities. In September 2002 a navy ship was anchored for two weeks outside Pasuruan. From claims made by prahu sailors, it appears that naval patrols had made good use of this time to obtain what they euphemistically referred to as 'cigarette money'. One Madurese timber merchant claimed that during this

---

[23] No reader should doubt the extent of such predatory corruption by police and the military in Indonesia. Prahus make easy targets because their owners and crews lack a public voice but truck drivers have also been frequent victims of extortion from state security personnel. On 1 February 2001 hundreds of container truck drivers staged a strike to protest at having to pay up to Rp 500,000 to be allowed past police and army security posts in Aceh (*Jakarta Post*, 5 October 2002: 3).

Photo 31 Unloading timber from a prahu to a truck, Pasuruan, 2002.

period two vessels supplying his warehouse had both been intercepted twice, first by an water police patrol and then by a naval patrol. In both cases, the water police patrol had demanded Rp 500,000 in return for allowing the vessels through, while the naval patrol had demanded and received Rp 500,000 from one vessel and Rp 3 million from the other. The merchant concerned was angry at his skipper for handing over so large a sum as Rp 3 million but he understood that the skipper had been afraid. He was also irate about the involvement of the navy, which he regarded as acting outside its jurisdiction. The same merchant claimed that he had also been subjected to extortionate demands for technical-legal shortcomings on the SKSHH, such as when a vendor in Kalimantan inadvertently gave the name of another timber merchant to the officials making out the permit. The fact that the name of the purchaser was wrongly stated on the permit was really irrelevant in this case because the name of the carrying vessel was correctly shown. Nevertheless, the merchant had to pay Rp 7 million to police in the port of Pasuruan before he could take his cargo.

Another problem for timber consignees, and one of critical importance, was the time factor. Each SKSHH is valid for only five days from the date of issue, which is clearly shown on the permit (Appendix 2). The vessel must reach the port stated on the permit within the specified time frame. The rationale for this limited time validity is to prevent the same permit from being used fraudulently to carry a second cargo. If the stipulated period was one month, for example, it would certainly be possible for a second cargo to be carried in that time, with opportunity for evasion of substantial revenue payments to the central government. In practice, however, two full voyages in five days would not be remotely possible. A longer period of validity, such as two weeks, would be more appropriate, especially as many timber-carrying prahus have rather small engines. Indeed, according to the Pelayaran Rakyat regulations in 2003, the engine was supposed to be only an auxiliary means of propulsion. Some of the smaller timber carriers were slow even in smooth water, let alone against a heavy head sea. In July and August, when the southeast monsoon is at its height, vessels leaving Kalimantan for East Java often have to struggle against strong headwinds and rough seas, making it difficult for some to reach their destination before the permit has expired. Similarly, any mechanical mishap can mean not making port in time. In cases of late arrival, timber merchants claimed that police would take advantage of the situation by demanding a heavy bribe before allowing the vessel to unload.

The problem of the period of validity of the transport permit was less pressing at Sepulu, which from southern Kalimantan was quicker by a full day than Pasuruan. However, by 2002 Sepulu was finally being targeted aggressively by the authorities, who now regarded it as the major conduit for illegal timber. Whether because of increased pressure from the upper echelons of the administration or because of less corruptible senior police, the old system of organised pay-offs was no longer sufficient to ward off an official crackdown. As at Pasuruan, the navy made its presence felt by stationing a vessel off Sepulu for a short period, escorting seven Sepulu vessels with no SKSHH for their cargoes to Surabaya. These vessels had apparently all paid the local *retribusi* in Central Kalimantan but not the central government taxes or royalties, without which an SKSHH cannot be issued.

At Sepulu partial compliance with the law was no longer a realistic option, the official stance in this respect having become even tougher than in Pasuruan. Forestry and police officials had become aware of the degree of evasion of taxes on timber imports and there were some among them who had been involved in actual cargo checks, involving unloading the whole cargo on to the quay. On the basis of such experience they could make a good estimate of how much a vessel might really be carrying. Numerous raids were conducted by police teams at Sepulu and nearby Telaga Biru and harsh on-the-spot penalties meted out on all suspect incoming cargoes. Refusal to pay would result in the vessel being towed or escorted to Surabaya, where the cargo would be unloaded and checked against the documentation. Quite apart from any penalty resulting from a subsequent prosecution, such a procedure would entail significant disruption of business and financial loss.

So successful was this campaign to enforce the timber transport regulations in Sepulu that by October 2002 virtually all shipments of timber to the port were accompanied by documentation for 90 per cent of the total amount carried, a level of compliance much higher than in Pasuruan. The harbourmaster was no longer obliged to 'go along with the culture', and any vessel without a timber transport permit was not allowed into the port. However, with the higher costs of compliance, the profitability of the timber business at Sepulu was slashed. Haji Komar summed up the general mood in the town by declaring gloomily, "In the past the timber business here was great but Law 41/1999 [the revised Basic Forestry Law] is fatal for us." Yet, despite this drastic decline in profitability, the volume of timber landed at the beach port abated only slightly. In late 2002, after several months of almost full compliance with the timber transport regulations, Sepulu was still a bustling place with teams of labourers busy throughout the day unloading timber from lightering vessels and carrying planks on their shoulders to the stacks piled along the beach, and 75 to 100 fully laden trucks rumbling off each afternoon and evening, bound for Surabaya.

The key to this continuing timber traffic despite the strict enforcement of the law was the Sepulu Entrepreneurs' Association. The leading figures of this organisation, Haji Komar and his counterpart in the desa of Prancak, Haji Rofi'i (the son of the previous incumbent, Nur Habib), realised that the only hope for Sepulu was for the local timber traders to stand together, all selling their timber for the same price according to the quality and species. Although it was necessary to compete against timber merchants elsewhere, it was vital to not attempt to undercut the local market. Of course, no merchant paying for an SKSHH on the full volume of his timber shipment would want to sell too cheaply for a modest profit margin had to be maintained. Nevertheless, unlike in the major mainland ports, including Pasuruan, port forestry and police personnel were not stationed permanently at Sepulu. Consequently, it was possible that whenever the coast seemed safe from a taskforce raid, some merchants might be tempted to bring in timber with an SKSHH for perhaps only 30 per cent of the total carried. This cheaper timber could then easily undercut the main market. In such a situation, local backers would be unwilling to invest in any venture with a high degree of legal compliance, knowing that their more expensive timber would not sell for a fair return. At the same time, few would dare to take the risk of backing illegal imports, with the possibility of losing everything if the timing went wrong. The volume of timber passing through Sepulu would inevitably decline, along with the port's reputation as a reliable source of timber. Buyers would turn to other places.

Despite the Madurese penchant for individualistic action, through the influence of the Sepulu Entrepreneurs' Association and especially its two leading figures, Haji Komar and Haji Rofi'i, there was no price-cutting competition at Sepulu. If a merchant was prepared to run the risk of only 30 per cent compliance, that would be his own business but the timber could not be sold for less than the prevailing price. In practice, and no doubt partly because of this strong social pressure to not undercut the market, virtually all incoming shipments did have an SKSHH for 90 per cent of the total carried. Although profits were now lean indeed by comparison with before, through belt-tightening and a continuing high turnover, the timber traders of Sepulu were able to remain in business.

The situation at Telaga Biru was very different. In September 2001 the port had still been a hive of activity, with several vessels unloading simultaneously and the warehouses lining both sides of the street along the foreshore filled with timber. A year later the place was moribund. All but one of the warehouses were locked up and through their barred paneless windows it could be seen that they were empty. The harbour itself presented a similarly forlorn spectacle. There were as usual many vessels, some of them large and in apparently sound condition, but now they were nearly all lying idle, some at anchor, others tied up to the outer breakwater. The one exception, the only vessel in port laden with timber, was a 'foreigner', a Bugis *pinisi* with its cargo being unloaded into the sole warehouse that was open.

Otherwise the port area seemed lifeless, with hardly a bicycle on the foreshore street in the middle of the day. The contrast with the bustle and vitality of Sepulu, only a few kilometres away, was stark.

That this impression of Telaga Biru in late 2002 was not deceptive became clear from conversations with locals. For decades there had been plenty of work for able-bodied men unloading timber. Now there was virtually none. Unemployment was rife and there did not seem any prospect of this changing. At the level of the sub-district (Tanjung Bumi), the main hope for the future was now from an oil extraction project about 20 kilometres offshore. The oil rig, clearly visible from the coast, had recently moved into a productive phase and a considerable amount of oil had already been landed at Telaga Biru. Output was expected to expand over the next few years, which would benefit the administration as well as provide a continuing justification for the well-staffed Telaga Biru harbourmaster's office. However, for the local men, most of whom lacked the skills in demand on the oil rig or the supply vessels, prospects of employment through this oil project were dismal. Although the cattle trade to Kalimantan was still viable, the other 'traditional' area of maritime enterprise from Telaga Biru, the passenger trade, was not. After operating in flagrant violation of the shipping regulations for decades, this trade had finally been closed down in the late 1990s, with prospective migrants to Kalimantan thereafter having to take a modern passenger ship from Surabaya.

For local entrepreneurs, the main prospects for Telaga Biru were now seen to lie not in maritime enterprise at all but in the fine batik produced by the women of Telaga Biru. Although batik is not made anywhere else in Madura, in Telaga Biru it is a long-established cottage industry. An oft-heard explanation of this phenomenon, given by men, is that the women of Telaga Biru turned to making batik so that they would not feel lonely, and perhaps take on lovers, while their husbands were away at sea. However, in many other maritime villages around Madura, many of the men would be away for the entire sailing season without any local batik or similar industry developing as a result. Nor would the absence of husbands during the sailing season necessarily have been as hard on their wives as might be imagined, because village life is socially quite rich and with much interaction between households. Probably it was the men on the prahus who were the more deprived by the lengthy separations. Batik making may have been introduced to Telaga Biru, or popularised there, by women from

Pekalongan in Central Java, who married prahu sailors and moved to Madura. *Golekan* from Telaga Biru were in fact regular callers at Pekalongan, a stronghold of the batik industry, and some of the prahu sailors did marry women from there (Chapter 6). In recent years the reputation of batik from Telaga Biru has grown, for the work is still done in the traditional way using a *canting* (pen), whereas most of the batik from Central Java is now semi-mass produced, using stamps. Yet, although some women in Telaga Biru have managed to provide good incomes for their families from batik, the niche market for this quite expensive cloth is far smaller than that for the high-quality timbers in which the traders of Telaga Biru had specialised, with correspondingly limited benefits for the community as a whole.

The reason for the economic malaise affecting Telaga Biru was both simple and predictable. As one official in the harbourmaster's office expressed it, "The difference between Telaga Biru and Sepulu is that there is no Association here." In the absence of such an organisation, such as that at Sepulu, the old Madurese pattern of individualism continued to prevail. No one at Telaga Biru wanted to pay for a 90 per cent compliance with the transport regulations. A few managed to bring in timber cheaply without being caught, but many did get caught until eventually no one was prepared to take the risk. By late 2002 there were only four or five vessels from Telaga Biru left in the timber trade. These were all large vessels, carrying 300–400 cubic metres of timber, at which size the economy of scale allowed a profit to be made despite full compliance with the timber transport regulations. These large vessels were moreover able to earn from both legs of each round trip, carrying livestock on the outward voyage.[24]

However, despite their continuing involvement in the timber trade, these large vessels of Telaga Biru no longer discharged in their home port. Instead they took their cargoes of timber, obtained from either Pontianak or Samarinda (in East Kalimantan), directly to Benoa, in Bali. Although Benoa is the main port of Bali and close to the capital of Denpasar, it had formerly been insignificant as a receiving centre for timber. Partly this was because of geography: Benoa is located in the southern extremity of the island and access to the port

[24] The cattle trade was still thriving. According to staff at the harbourmaster's office, interviewed in September 2002, around 1000 head of cattle were being exported per month.

through the Lombok Strait is complicated by strong currents and, during the height of the southeast monsoon, strong headwinds. For these reasons the main timber port in Bali used to be Celuk Bawang, near Singaraja on the north coast. Celuk Bawang is sheltered from the southeast wind and prahus can lie comfortably to anchor while their timber is ferried to the beach. After the almost universal fitting of engines in timber-carrying prahus, the problem of accessibility to Benoa no longer applied. Nevertheless, most prahus continued to use Celuk Bawang. This was not only because it was closer and more convenient but, more importantly, because the timber transport regulations were not enforced there. After the general crackdown on timber smuggling from 2001 onwards, the laxity prevailing in Celuk Bawang came to an end and the larger vessels turned to Benoa, where a higher price could be obtained.

The era of timber trading in complete disregard of the timber transport regulations thereby came to an end. Starting with the prahu ports of Jakarta, the requirement that timber cargoes be accompanied by an appropriate transport permit had gradually been enforced right along the north coast of Java and Bali. Timber merchants had in general adjusted to this policing regime, with price increases passed on to consumers. In East Java, however, the transition to compliance with the law had been less smooth than in West Java and Central Java. Surabaya and Gresik had been the first two ports to be closed to illegal cargoes, followed by Sedayu Lawas, Pasuruan and finally the timber ports of Madura. In Gresik and Surabaya, enforcement of the regulations had made little difference to the volume of traffic, because the demand was so high. In the case of Sedayu Lawas, this prahu harbour of old had recovered as a timber port serving the western portion of East Java and been given improved facilities and a harbour-master's office. It was now used only by medium-sized Bugis vessels, whereas in the days when its inlet had been crammed with small Madurese prahus, Bugis vessels had never called there. In Pasuruan, the traffic of prahus had at first been seriously disrupted by attempts to enforce the law, but the future of this important prahu port had never been in doubt because it served such a large and heavily populated hinterland. The timber ports of Madura were much more precarious. Telaga Biru appeared virtually finished as a timber port, while Batioh and Pagar Batu had suffered a similar fate. Only in Sepulu had the crisis been weathered, albeit with difficulty.

5

Incomes and Profits

Although the previous chapter described the role of the maritime Madurese in the domestic timber trade up to 2003, it did not touch upon the important subject of incomes and profit-sharing arrangements. This chapter will present three case studies of Madurese involvement in the Kalimantan-Java timber trade during 2002–03. Apart from showing earnings and profits from voyages, these case studies allow more information on ports and routes than has been possible so far, as well as more detailed explanations of the workings of the timber trade at both the loading and destination ports. The case studies also provide opportunity for richer ethnographic description. The chapter concludes by noting connections between Islam and commerce in Madura.

At the outset, it is useful to clarify 'profit sharing'. Two sorts of profits will be dealt with in this chapter: profits from carrying timber and profits from trading in timber. In the former, the vessel earns income for each cubic metre of timber carried. After operating costs have been deducted, the net vessel income is divided into a number of equal shares, which are then parcelled out among the stakeholders in varying proportions according to the arrangement concerned. These stakeholders include the vessel owner, the engine owner (sometimes the vessel owner, sometimes not) and the crew, with the skipper a special category among the latter. The crew thus share in the voyage profits rather than being paid in wages. The one significant exception up to 2002–03 was at Sepulu, where the crew were paid in wages. These fixed rate wages were per voyage, not per day, so the crew still had incentive to complete each voyage as quickly as possible. This arrangement was adopted to keep vessel operating costs low, thereby assisting the Sepulu timber traders to undercut the market in Pasuruan and Surabaya. Although the earnings under this wage system were much lower per voyage than earnings through profit-sharing, the Sepulu sailors were making up to four voyages per month compared to just one voyage per month elsewhere.

In the case of trading, there is no profit sharing because there is usually only one direct stakeholder — the person who funds the purchase of the timber in Kalimantan. Except in the case of large vessels carrying 400 cubic metres or more, at which size freight returns can be good, most Madurese vessel owners aspire to become involved as traders as well as carriers of timber. However, most vessel owners lack sufficient capital to become traders. More often than not, the person funding the timber purchase is either a timber merchant in Java or Madura or an outside investor with no direct involvement in the buying and selling. Examples in the case studies should clarify these differences. Investors in timber voyages do not necessarily need sufficient capital in their own right to pay for a full cargo. As in the case of Haji Ansuri mentioned in the previous chapter, funds are often borrowed, in the form of gold, from others in the community.

Of the three following case studies, the first concerns vessels from Sumenep (Giligenting and Sapudi) with emphasis on the Sukamara (Central Kalimantan) to Juwana route; the second concerns vessels from Sreseh, with emphasis on the Kintap (South Kalimantan) to Pasuruan route; while the third focuses on vessels from Sepulu, carrying timber from Sungai Kaki, a small river in the southeast corner of Central Kalimantan, to Sepulu. These are by no means the only timber routes used by Madurese vessels but they typify Madurese involvement in the timber trade during 2002–03. In addition to presenting information on incomes and profits, ports and routes, and the quasi-legal nature of the contemporary timber trade, these case studies provide further evidence of the mobility and flexibility of the maritime Madurese in the face of changing economic circumstances.

Sukamara to Juwana: The Importance of the Agent

Although prahus from Telaga Biru and Sreseh used to be common callers to Central Java, more recently all Madurese prahus discharging timber in Central Java have been from Sumenep. They use three ports, from west to east Tegal, Semarang and Juwana (see Map 12, Chapter 4). Semarang is a modern port used by ships as well as prahus and the Madurese vessels calling there tend to be larger on average than those calling at Tegal and Juwana. Some of the Madurese vessels discharging timber at Semarang also carry outward cargoes of goods such as rice and fertiliser. However, for small- to medium-sized Madurese prahus, the ports of preference are Tegal and Juwana, mainly because port costs in those places are lower than in Semarang. Like Pasuruan, Tegal and Juwana are too shallow to allow ships to enter and have long been used by only prahus. Both ports are of long standing, with Juwana especially important in the eighteenth century. Juwana is situated close to Rembang, formerly the main centre for indigenous shipping (see Map 6, Chapter 1). The port is located on a river, whereas Rembang itself lacks a natural harbour. In both Juwana and Tegal, timber is almost the only inward cargo, while very little cargo is shipped outward. Although the town of Juwana has a smaller population than either Tegal or Semarang, as a destination for timber-carrying prahus Juwana has the advantage of being the closest of all ports in Java to the coastline of Central Kalimantan. This is significant because in both South Kalimantan and especially West Kalimantan the forest resources have been heavily exploited and much clearing has taken place for agriculture, leaving Central Kalimantan as much the most important province to supply the Java timber market.

Although this case study focuses on Juwana, there is no significant difference in the timber trade there and at Semarang and Tegal. However, until at least 2003 there was an important difference between the timber trade in the ports of Central Java by comparison with those of both West Java and East Java: the law concerning the movement of timber was not enforced as rigidly in Central Java as in the two neighbouring provinces. Raids by police and forestry teams did occur in Central Java but only very infrequently because of a system of organised graft payments to the parties concerned. These graft payments, euphemistically referred to in the trade as 'coordination money', were made by agents to various authorities as a normal part of the processing procedure for each shipment. Although such 'coordination' payments did not obviate the need for vessels to have timber transport permits for their cargoes upon arrival in port, the standard practice in Juwana, Semarang and Tegal was that the SKSHH covered only 50 per cent of the actual cargo. A similarly low degree of compliance was still common in Pasuruan in late 2002 but, as explained in the previous chapter, the situation there was uncertain for non-complying timber consignees. Moreover, vessels coming into

Pasuruan were being plagued by demands for payoffs to water police and navy patrols. Vessels discharging timber in the three Central Java ports were untroubled by these problems. As a shipping agent in Juwana explained:

> Juwana is a community port. All the officials around here are locals. It's all worked out in advance... everyone gets paid off, the DPC [Dewan Perwakilan Cabang, local branch council of Pelra], police, forestry. There is hardly ever any trouble with raids. No one wants that. It all depends on the local community — when everyone cooperates, everything goes smoothly.

Judging from my conversations with officials and port workers, there did indeed seem to be a relaxed approach to business at Juwana, with none of the underlying tension encountered at Pasuruan. In the case of low-grade timber for use in rough and temporary construction work, even the 50 per cent level of compliance was often not bothered with. A problem for entrepreneurs buying low-grade timber in Kalimantan and selling it in Java was that while such timber naturally sold on the market for a lower price than would be obtainable for better-grade pieces of the same species, both national and district timber taxes were imposed regardless of quality and purchase price. Consequently, buyers of low-quality timber felt that they could not make an adequate profit if they paid for an SKSHH, even at only 50 per cent compliance. Thus in 2003 shipments of low-grade timber were still being landed completely illegally in Central Java, with the associated risks minimised by the system of graft payments.

Most of these illicit shipments came from the Central Kalimantan town of Samuda, just inside the entrance to the Mentaya river, and were being carried by Madurese prahus. Samuda had formerly been a predominantly Madurese settlement but there had been good relations with the indigenous Dayak community. In 2002, a year after the ethnic strife centred on Sampit (see Chapter 4), Madurese vessels had resumed calling to Samuda but not to anywhere else along the Mentaya. In fact, most of the vessels loading at Samuda in early 2003 were Madurese. The original advantage of Samuda, its accessibility under sail, was no longer important after motorisation, while the timber available there was of poorer quality than further upriver. However, Samuda had the great advantage, by virtue of its location near the river entrance and its distance from Sampit, that there was less chance of a forestry official or police patrol coming on board and demanding to see a transport permit for the cargo. In fact, an official investigation in late 2002 into the volume of the timber trade on the Mentaya concluded that the majority of the vessels carrying timber from Samuda were departing for Java without any documentation whatsoever, not even a port clearance. A senior port official at Samuda told the investigative team that he refused to issue port clearances to the vessels concerned because they were carrying baulk timber for which no district tax had been paid. Clearly, the refusal to issue port clearances had no impact on the trade (*Borneo*, 48, 2002: 4). Yet although such illegal shipments of low grade timber were still common in 2003, timber coming into Central Java was mostly of medium grade or better, with a 50 per cent level of documentation as noted above.

The crucial figure in this system of quasi-legal timber importation was the shipping agent. Since 1964, under a law reorganising the shipping industry (Law 5 of 1964), all port administrative business had been required to be conducted through an authorised shipping agent registered as a limited liability company (*Perseroan Terbatas*, PT) (Dick, 1975a: 82). Some agents also carried out a forwarding function, arranging cargoes for larger vessels. In the context of the timber trade, the facilitative function of the agent had expanded greatly, so that many shipping agents were now pivotal players not only in the prahu shipping industry but also the timber trade itself. A case in point in 2003 concerned an entrepreneur from Nipa, the desa adjoining Batioh, on the north coast of Madura. This man, Haji Sahid, had become involved in bringing timber from Central Kalimantan to Pagar Batu at the eastern end of Madura. He had no vessel of his own and used an engineless vessel from Giligenting named *Dharma Bhakti* (Loyal Service), owned by a man named Pak Sum. Business had been going very well for Haji Sahid until late in 2001 when *Dharma Bhakti*, carrying 100 cubic metres of timber from Kumai with no documentation at all, was intercepted by a water police patrol off Kalianget. As the owner of the timber, Haji Sahid was arrested and brought to Sumenep, where he was fined Rp 38 million and sentenced to three months in prison. His total losses were much greater than that, however, for the entire cargo was confiscated. After this incident, Haji Sahid dared not resume bringing shipments of undocumented timber into Madura, whether on the northwest coast or at Pagar Batu. Yet if he imported fully legal timber, his profit margins would be negligible. However, he knew that in Central Java incoming vessels were not being harassed in the same way as in East Java. As it happened, he knew of a man from Telaga Biru, not far to the west of Batioh, who had a shipping agency in the port of Juwana. He contacted this agent, a man named Abdul Majid, who confirmed that he could arrange consignments of timber for *Dharma Bhakti*, with Haji Sahid acting as financial backer.

To start bringing in timber to an unfamiliar area without specialised knowledge and local contacts could have been risky and fraught with administrative, legal and financial problems. Haji Sahid had no such worries, however, because virtually everything to do with the actual buying and selling of the timber was organised by his agent in Juwana. Given the large amounts of money involved, the absence of any written contract between the parties concerned and no surety against loss, it is obvious how critical was personal trust in such a business relationship. In this case, the fact that Haji Sahid's agent, Abdul Majid, was a fellow Madurese was very important because Madurese usually trust their ethnic fellows in preference to agents of other ethnicity whom they fear might take advantage of their lack of specialised knowledge. Indeed, it would be easy for an agent to increase his profit through collusion with timber merchants and by inflating the various official and unofficial charges incurred in shipping timber. However, agents acting so unscrupulously would quickly lose both their reputations and their clients. Many financial backers have an imperfect understanding of the procedures involved in the timber trade but they can and do compare their returns with the experiences of others.

Abdul Majid's background in Telaga Biru stood him in good stead as an agent in Juwana because, after the decline of the timber trade at Telaga Biru and Batioh, many entrepreneurs from northwest Madura were keen to invest in the trade to Central Java. It was not necessary for a prospective investor to own a prahu or have an arrangement with a prahu owner, as in the case of Haji Sahid and Pak Sum. Abdul Majid could easily organise a vessel because the increase in size of the vessels working the timber trade meant that many older but still serviceable smaller prahus had no work. By 2003 Abdul Majid had ten financial backers for the timber trade through Juwana, all of them from Telaga Biru or thereabouts. At least one of these Telaga Biru entrepreneurs was also involved in exporting cattle from Juwana to Pontianak in West Kalimantan, with Abdul Majid organising the shipments from the Juwana end on behalf of his client. The receiving cattle merchant in Pontianak was also a native of Telaga Biru.[1] In all cases the vessels concerned were from either Giligenting or Sapudi.

Until his arrest and subsequent imprisonment and confiscation of his cargo in late 2001, Haji Sahid had been obtaining his timber from the port of Kumai (see Map 13, Chapter 4). This was a convenient port for an engineless vessel like *Dharma Bhakti* because the Kumai inlet is broad and the port easily reached under sail. For this reason Kumai had long been popular with Madurese prahu skippers, many of whose vessels were still relying on sails two decades after the Bugis fleet had been motorised. Kumai itself is a small place but only 11 kilometres to the large town of Pangkalanbun, which until 2001 had a strong Madurese presence. Pangkalanbun was originally reached via the Kotawaringin river, on which the town is situated. However the Kotawaringin is only a narrow waterway and as the town grew Kumai became the principal port, with even large interisland passenger ships calling there.

While the Kumai river is broad, it is not a long arterial waterway like the Mentaya and all the land around has been cleared of forest. Consequently, timber is brought to the port by trucks, using a network of roads extending deep into the hinterland.[2] Although the timber being delivered to Kumai was still of good quality, by 2002 it was taking longer than before to obtain cargoes there because logging was taking place so far into the interior. Instead, Haji Sahid's agent, Abdul Majid, recommended Sukamara, about 80 kilometres away in a straight line to the west of Kumai. Sukamara is situated on the east bank of the Jelai river, which forms the provincial boundary: the east bank of the river is in Central Kalimantan while the west bank

---

[1] I saw the documentation and costing for one such shipment of 125 cattle. Freight was Rp 125,000 per head, on top of which the exporter had to pay port and agent costs of Rp 25,000 per head. Significantly, the 'port' costs included a payment to the local water police of Rp 200,000, even though the police had no administrative involvement.

[2] In 2003 one could see rafts of logs at Kumai but they were logs of *ramin* (*Gonystylus* sp.) taken illegally from the nearby Tanjung Puting nature reserve. Ramin is in high demand on the international market because its very clean grain makes it ideal for uses such as picture framing. The prahus serving the domestic timber market have no involvement in this trade, these ramin logs being carried on large steel barges to Singapore and Malaysia (EIA, 2001; *Tempo*, 29 July 2001: 68).

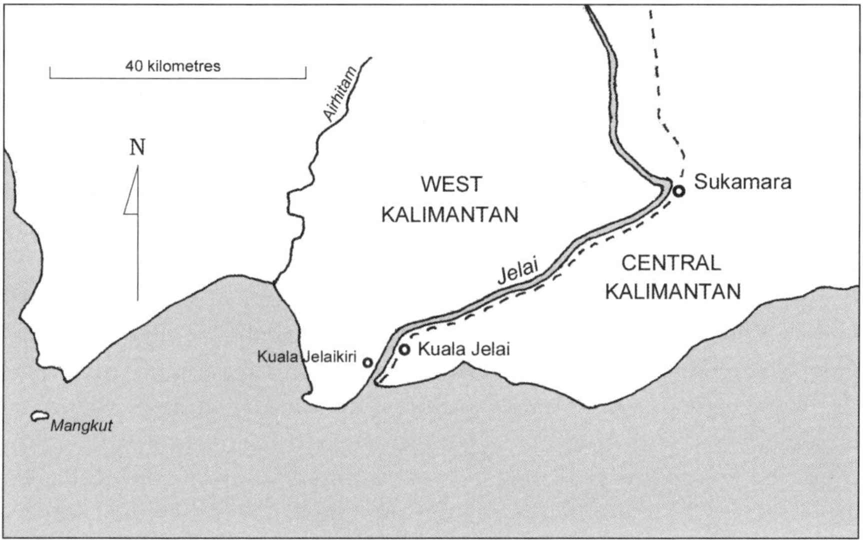

Map 15 The Jelai river and surrounds

is in West Kalimantan (Map 15). Although about 60 kilometres from the river entrance, Sukamara is not too difficult for engineless prahus to reach because the river is broad and fairly straight as far as Sukamara. In addition, the orientation of the lower part of the Jelai is optimal for sailing vessels, enabling them to take the prevailing wind on the beam in both directions. In any case, *Dharma Bhakti*, having started her career rigged as a *lete-lete*, had been converted in the 1990s to a sloop rig, making tacking easier. Nor was navigation any problem: prahus bound for the Jelai river made first for the promontory of Tanjung Puting (see Map 13), which has a light, then made their way to the northwest in daylight until they reached the river entrance. Being wide, the entrance is easy to see even though the land around is all low-lying. If a prahu accidentally went too far to the west, the crew would soon become aware of their error when the small island of Mangkut came into view. Mangkut was familiar to many older Madurese sailors, as being situated at the southwest tip of Borneo it used to be an important 'signpost' for prahus laden with salt or cattle bound for West Kalimantan.

Being well removed from the main centres of development in Central Kalimantan, the forest adjacent to the Jelai had been less heavily exploited than that along the Mentaya or in the hinterland of

Kumai. Nevertheless, even before motorisation some Sumenep prahus used to load timber in the Jelai, at Kuala Jelai just inside the entrance, or Kuala Jelaikiri on the western shore. By 2003 there were numerous sawmills up to and beyond Sukamara and the logging had taken its toll on the easily accessible areas of forest. The felling area was then about 50 kilometres inland from the town, whereas it was claimed that in the early years of the timber boom the felling area had been only about five kilometres from the river.

By early 2003 Sukamara had become the port of choice for many prahus from Sumenep because of the ready availability of timber. The only Madurese prahus then using Kumai were large vessels that were also carrying outward cargoes from Java. Some Madurese prahus at Sukamara were still loading cargoes of squared logs, trimmed roughly from the round by chainsaw to enable them to stow well. Squared logs used to be particularly suitable for small sawmills with limited equipment, even pit saws operated by hand. After the proliferation of powerful mechanised saws, much less baulk timber was exported to Java and virtually none landed in either of the Jakarta prahu harbours. Nevertheless, squared logs were still in demand, especially durable species such as bangkirai or benuas, for which species baulks were fetching up to Rp 2 million per cubic metre in 2003. For a sawmiller, squared logs could still bring in a good profit because sawmillers made money not only by sawing logs into planks but also by trading in timber. They bought their logs from local logging entrepreneurs and sold the wood with their own trading margin, in addition to processing costs. Moreover, selling as a baulk was a good way of dealing with small-diameter logs because even though such small logs did not yield wide planks, there was always a demand for solid pieces suitable for heavy posts. While most Madurese prahus carrying timber from Sukamara had only a few pieces of baulk timber stowed on deck, in some cases the entire cargo consisted of squared logs 30 to 40 cm square by 4.5 metres long, cut to this length to facilitate loading onto trucks when extracting them from the forest. In most cases these cargoes of squared logs were discharged at Tegal and thereafter transported to West Java. With timber being landed in Central Java at a discount rate because of limited compliance with the timber transport regulations, it was cheaper for West Java timber merchants to obtain their supplies through Tegal, just a short distance across the provincial boundary, than from Jakarta where the timber transport

Photo 32 Unloading squared logs from a large Madurese vessel at Tegal, 2003.

laws were more vigorously enforced. There were, however, risks involved in doing business like this, as will be explained in the second case study.

Some vessels were still sailing from Sukamara in 2003 with no documentation at all for their cargoes. This was not because of laxity on the part of local officials. Until the second half of 2002 Sukamara was part of the administrative district of Kotawaringin Barat. Then a separate administrative district of Sukamara was formed, with affiliation to the 'mother district' of Kotawaringin Barat. After this reorganisation, officials in Sukamara were well aware that their salaries depended, at least indirectly, on the revenue raised from timber being carried out of their district. However, their task of ensuring that district taxes had been paid for cargoes going down the Jelai was complicated by the fact that, from Sukamara to the sea, the river forms the boundary with West Kalimantan. On occasions when a 'team' had been carrying out spot checks on fully laden prahus for documentation, some prahus had managed to avoid scrutiny by relocating temporarily to the west bank of the river, under the jurisdiction of West Kalimantan. Nevertheless, a vessel leaving Sukamara without adequate documentation for its cargo was not safe until it had cleared

Kuala Jelai, at the entrance to the river. In early 2003 a vessel from Giligenting was arrested in Kuala Jelai for having no documentation for a full load of bangkirai and benuas timber. Six months later the case had still not been resolved, a disaster for both the vessel owner and the financial backer for the cargo.

After his own calamity in 2001, culminating in his imprisonment, Haji Sahid had no desire to attempt to evade revenue payments altogether on his timber shipments. Through the offices of his agent in Juwana, Abdul Majid, he was obtaining satisfying returns with minimal risk, financing shipments of sawn timber from Sukamara. Haji Sahid was not directly involved in the buying of the timber and, indeed, he had never been to Sukamara. Instead, his shipments were arranged through Abdul Majid's Sukamara agent, who was also a native of Telaga Biru and in fact a cousin of Abdul Majid, again facilitating the vital element of trust.

To purchase timber in Kalimantan and bring it to Java involved three main areas of expense: the cost of the timber, documentation and associated charges, and freight. In 2003 Haji Sahid was paying Rp 350,000 per cubic metre for timber in Sukamara. With his carrying vessel, *Dharma Bhakti*, having a capacity of 100 cubic metres, the total cost of his timber for one shipment was thus Rp 35 million. However, he also had to pay for the timber transport permit, the SKSHH. This was a substantial amount, mainly because of the forest royalty fee (PSDH) which had to be paid first to qualify for the SKSHH. For the most common timber, meranti, the cost of obtaining an SKSHH was said in 2003 to be Rp 70,000 per cubic metre if paid directly by a prahu skipper to the local forestry office. In practice, however, it was usually paid through an agent for an additional fee of about Rp 20,000 per cubic metre. In the case of timber bound for Juwana, the SKSHH was routinely paid on only 50 per cent of the cargo, the balance being illegal. This level of compliance with the law was what the market in Juwana would bear, it being common knowledge in the trade there that a shipment with an SKSHH for 100 per cent of the amount carried would not fetch any higher a price on the market than a shipment with only 50 per cent compliance. In Haji Sahid's case, the SKSHH for his half-legal shipment came to Rp 4.5 million. This was not the end of his documentation costs, however, because he also had to pay a district timber tax in Sukamara of Rp 50,000 per cubic metre. This was not necessary for the consignment to be legal in Java, but officials in Sukamara would not issue the SKSHH without payment of the district timber tax. Again, this was paid for only 50 per cent of the cargo, coming to Rp 2.5 million. Such deceit was made possible by the vessel's agent in Sukamara, who knew how to avoid arousing the suspicions of local officials.

The other main expense in bringing timber from Kalimantan to Java was the freight charge, payable on arrival of the vessel in Java. In 2003 the rate for a sailing prahu from Sukamara to Juwana was Rp 90,000 per cubic metre. Haji Sahid therefore had to pay Rp 9 million for the transport of his 100 cubic metre shipment. Here he was making a significant saving because the freight cost for a motor vessel of the same capacity would have been Rp 11 million. In the event of *Dharma Bhakti* having to take a tow upriver, as could happen towards the end of the southeast monsoon when the weather became fickle, the cost of this would be borne by the vessel's owner. Haji Sahid was also liable for the cost of the agent in Sukamara: at Rp 15,000 per cubic metre, this came to Rp 1.5 million. The cost of the agent at the Juwana end was the responsibility of *Dharma Bhakti*'s owner. Finally, Haji Sahid had to pay a 'coordination' charge of Rp 1 million to his agent in Juwana so that the shipment would be allowed through without undue official scrutiny.

With the documentation all in order, the cargo was sold as a single lot, 'free on board', to a timber merchant in Juwana, the costs of stevedoring and subsequent administrative work thereby becoming the responsibility of the receiving merchant. The following is a breakdown of costs for Haji Sahid to bring a shipment of the cheapest and most common timber, mixed meranti, to Juwana in June 2003 (all figures in rupiah per cubic metre; Rp 10,000 = approximately US$1).

Purchase price (m^3)	350,000	(paid to sawmiller)
Freight charge	90,000	(110,000 if motor vessel)
SKSHH	45,000	(90,000/m^3, 50% compliance)
District timber tax	25,000	(50,000/m^3, 50% compliance)
Agent fees (Sukamara)	15,000	
Coordination costs	10,000	(Rp 1 million for shipment)
Total outlay	535,000	
Sold on board	615,000	
Profit margin	80,000	

On his 100 cubic metre shipment, Haji Sahid was thus making a net profit of around Rp 8 million, a 15 per cent return on his capital outlay over a period of just one month. If the timber was of good quality, his profit could have been even higher because Abdul Majid always selected the best pieces — about 20 per cent of the total landed through his agency — for the lucrative foreign export market. He would arrange for this top-quality timber to be sold to a local dealer at a higher price than would otherwise be obtainable in Juwana and this dealer would send the timber off in full truckloads to another dealer in Jakarta, who handled the actual export. To be acceptable for export, pieces had to be at least four metres long, of fairly substantial sawn size, usually six centimetres thick by a minimum width of fifteen centimetres, and free of flaws. Flawlessness was the prime factor on the international market, not the species of timber.

Given that such profits could be made from buying timber in Kalimantan and selling the whole shipment on arrival in Java, the question naturally arises as to why more Java-based timber traders were not themselves involved as financial backers for voyages to supply their own businesses. Some Java-based timber merchants did in fact purchase their timber from Kalimantan and have it shipped to Java but for various reasons many others bought 'free on board' instead. The main reason for buying free on board was the considerable element of risk associated with buying and transporting timber from Kalimantan, especially if the shipment was not fully legal. Buying a cargo as a single lot on board in Java eliminated this risk. Second, the profits made from selling a cargo subsequently within Java were usually greater than those that could be obtained from buying in Kalimantan and selling free on board in port in Java. Third, many timber merchants lacked the capital needed to buy their timber directly from Kalimantan. In the absence of sufficient capital, there was ample opportunity for external backers to become involved as voyage financiers. Nevertheless, the sale of the cargo free on board was not left to the vagaries of the market. Instead, the sale would be arranged in advance, usually through a shipping agent with contacts in the timber trade. There was nothing remarkable about the intermediary role of prahu shipping agents in this respect because their businesses depended upon the timber trade.

In Juwana in 2003, the usual payment arrangement was that the receiving merchant would pay 50 per cent of the sale price initially, with the remainder being paid 4 to 6 weeks later after the timber had been sold to consumers. In 1999–2000, when the demand for timber had been strong, payment had usually been made in full. Nevertheless, having to wait for several weeks for payment to be completed did not necessarily slow business down, provided that the financial backer concerned had sufficient resources. To keep a vessel operating at maximum efficiency in 2003, a backer needed funds totalling three times the cost of a single shipment. Haji Sahid had such resources, thereby benefiting not only himself but also the vessel owner and the crew. Ultimately, of course, the viability of the timber shipping business depended on the construction industry. While the demand for timber fluctuated greatly between 1997 and 2002, it seems likely to remain fairly buoyant because of ongoing construction projects. From the perspective of those in the prahu shipping trade, the main problem was not the market but the large amount of capital needed for timber voyages.

The problem of capital for the financing of voyages had affected prahu owners, most of whom lacked the resources to become involved as timber traders in their own right. The worst affected were those whose prahus were without engines. Most investors in the timber shipping trade now preferred to use motor vessels, despite the higher freight costs involved, because they were regarded as more efficient. Yet to install an engine was very expensive and hardly worthwhile if the vessel was already old like *Dharma Bhakti*. This vessel's owner, Pak Sum of Giligenting, was therefore fortunate in his business arrangement with Haji Sahid, although the benefits went both ways because Haji Sahid was getting his timber transported at a considerably cheaper rate than would be the case with a motor vessel. Pak Sum's returns in 2003 were more modest than those obtained by his 'timber boss' but nonetheless adequate, as shown by the following simplified breakdown for a sailing prahu charging a freight rate of Rp 90,000 per cubic metre (all amounts in rupiah):

Gross freight earnings	9 million (cargo of 100 m^3 × 90,000)
Less operating expenses (agent fees, port fees, victualling)	2 million
Net freight earnings	7 million

In *Dharma Bhakti*'s case, the net freight figure was divided into ten equal shares of Rp 0.7 million, four shares for the owner and one share for each of the six crew members including the skipper. The skipper received more than this, however, because he was also paid a bonus by Pak Sum for each completed voyage, worth about half as much again as his standard share. After paying the skipper's bonus, the owner, Pak Sum, was therefore receiving around Rp 2.5 million for each voyage in 2003, amounting to about Rp 17 million for a seven voyage sailing season. He was liable for all maintenance costs but these were much less than for a motor vessel, with the mast made from a young tree, the sails made by the crew from cheap woven plastic cloth, the standing rigging of galvanised wire cable with no special marine fittings, the bilge pump homemade, and the hull below the waterline coated not with expensive anti-fouling paint but the traditional mixture of slaked limestone and oil. Barring serious mishap, Pak Sum could expect fair returns as long as his vessel remained serviceable and had steady work.

In the case of motorised vessels, the arrangement was different. The freight rate from Sukamara to Juwana in 2003 for a vessel with a four-cylinder engine was Rp 110,000 per cubic metre, while for a vessel with a more expensive six-cylinder engine the rate was higher at Rp 120,000 per cubic metre.[3] Voyage backers with sufficient means were generally prepared to pay the higher rate because a vessel powered with a six-cylinder engine would be faster than one with a smaller power unit. By modern motor vessel standards, many smaller prahus had insufficiently powerful engines to allow good progress in head seas, especially when heavily laden. The following simplified breakdown is for a motor vessel from Giligenteng with a six-cylinder engine, carrying 200 cubic metres of timber in 2003:

Gross freight earnings	24 million (cargo of 200 cubic metres)
Less operating expenses:	
Fuel	5.4 million (3,000 litres × 1,800)[4]
Agent fees, port fees, victuals	2.6 million
Net freight earnings	16 million

[3] Some vessels were also chartered directly for a fixed overall charge: one from Giligenting with a capacity of 200 cubic metres was being chartered in 2002 for Rp 30 million per voyage from Sukamara to Tegal. Arrangements for the vessel owner and crew nevertheless remained the same as for a vessel earning freight per cubic metre carried.

[4] Diesel fuel cost Rp 1,800–2,200 per litre in Indonesia in 2002–03.

This net freight figure was then divided into three equal shares: one share to the vessel owner, one share to the engine owner, and one share for the crew, usually six persons on a vessel of this size. Because the engine was so expensive, often as expensive as the vessel itself, it sometimes happened that a separate investor owned the engine. Usually, however, the vessel owner was also the engine owner. In the latter case, the owner would receive two-thirds of the net earnings (Rp 10.6 million in the above example). However, some Sumenep owners of both vessel and engine were said to take only 60 per cent rather than two-thirds of the net freight earnings. Assuming eight voyages a year, the owner of a vessel carrying 200 cubic metres of timber would have received a gross annual income in 2003 of about Rp 80 million. Given that a new vessel of this capacity could at that time be built and fitted out for between Rp 150 and 200 million, including purchase and fitting of the engine, the capital outlay could thus be recovered in less than three years. For those who could afford to build or buy vessels of this capacity or larger, maritime transport could therefore still be profitable.[5]

The crew share in Sumenep motor vessels was apportioned as follows. For a six-man complement, the share (one third, or sometimes 40 per cent of the net freight earnings) was divided into 13 equal lots, 2 each for ordinary crew members and 3 for the skipper. If the vessel had an engine-room boss (*Kepala Kamar Mesin*, KKM), as was often the case with larger vessels, the crew share would be divided into 14 equal lots, 2 for each ordinary crew member and 3 each for the skipper and the KKM. While having an engine-room boss reduced the proportion of the crew share available for ordinary crew members, this was more than compensated for by the larger size of vessels having such an arrangement, usually 300 cubic metres capacity or more. In 2003, on large motorised Madurese prahus calling to Juwana, Tegal or Semarang, ordinary crew members were typically receiving about Rp 1,000,000 per voyage, equating to about Rp 30,000 per day. This compared very well with ordinary labour rates in Java and Madura: for example, stevedoring labourers in Tegal and Juwana were receiving only Rp 10,000 to 15,000 per day on a casual basis.

[5] Motor vessels, especially moderately large ones, usually had longer working seasons than sailing vessels. Some of the largest motor prahus were working right through the northwest monsoon.

In addition to their income from the net freight division, skippers from Giligenting and Sapudi received a substantial bonus from the owner for each completed voyage to encourage honesty and fast turn-around in port. This bonus could bring a skipper's earnings per voyage up to three times that of his shipmates. Finally, there was a free timber transport allowance for crew members who wished to do some small-scale trading on their own behalf. Usually this allowance was one cubic metre per person but on a large vessel a skipper or engine-room boss might have an allowance of up to three cubic metres. On *Dharma Bhakti* this crew transport allowance was taken up in the form of a few baulks carried on top of the large hatch cover. These were the only baulks carried, the main cargo consisting entirely of planks. Costing between Rp 250,000 and 300,000 per cubic metre in 2003, these baulks could be sold privately in Java or back in Giligenting at the end of the sailing season for up to Rp 800,000 per cubic metre. If a crewmember lacked sufficient funds of his own, he could buy timber to the extent of his transport allowance on behalf of another party, usually a friend or family member, and later share the profit with that person.

Kintap to Pasuruan: Freewheeling Shipper-traders

By comparison with the ports of Central Java and even Gresik and Surabaya, the port of Pasuruan is visited by a great variety of prahus. Among the vessels discharging timber in Pasuruan in 2002–03 were Mandar prahus from Kota Baru in South Kalimantan, Bugis prahus from Makassar and Bima (Sumbawa), and Madurese prahus from Pasian on the northeast coast of Madura, Sreseh on the south coast, and Giligenting and and Sapudi. While vessels from Bima were the most common, all the above groups were well represented. Within the context of this study, however, the timber carriers and traders from Sreseh warrant special attention.

At Pasuruan, entrepreneurs from Sreseh were prominent in the timber trade. This may seem surprising, considering that Pasuruan is such a large and bustling town and Sreseh just a small and isolated place on the other side of the Madura Strait. However, the strong outward migration from the Sampang-Sreseh area historically focused on Pasuruan and nearby Probolinggo, whereas the outward migration from Pamekasan and Sumenep focused on Besuki, Situbondo and inland areas of the eastern portion of Java (Husson, 1997). Even today,

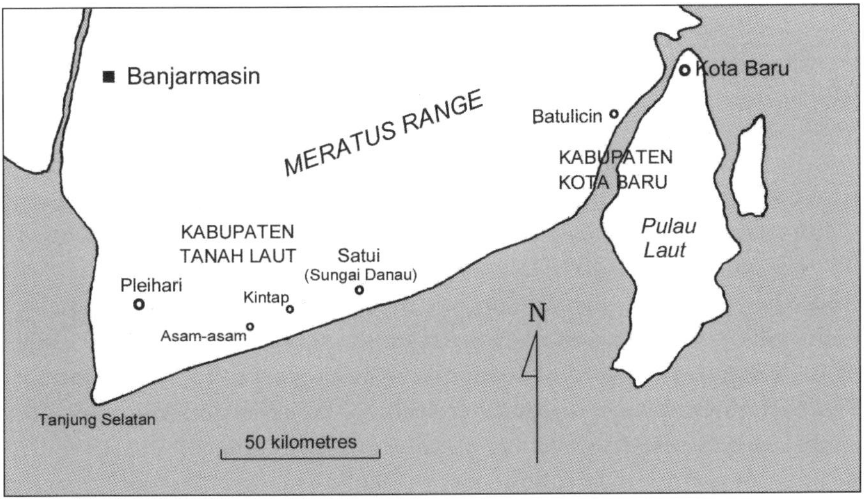

Map 16 Banjarmasin to Kota Baru

the orientation from Sreseh is more strongly towards Pasuruan than anywhere else in Java, with strong social ties between the two places. Moreover, salt from the Blega estuary adjoining Sreseh was transported by prahu mainly to Pasuruan, resulting in a strong maritime connection with that port (see Chapter 3). Indeed, some individuals from Sreseh who became involved in the transport and trading of salt to Pasuruan later moved into the timber trade in that port when the salt business began to decline during the 1980s.

One entrepreneur from Sreseh who followed this pattern was Haji Matsahri, of Batuputih, a prahu owner and also joint owner, together with his eldest son, Marthawi, of a timber wholesaling business in Pasuruan. Although he had retired a few years earlier from working as a prahu skipper, in 2002–03 Haji Matsahri was not much involved in the day-to-day running of the timber warehouse, which was being managed by his son Marthawi, who lived with his family on the premises. Haji Matsahri's role was instead the transport side of the business. He was living in Sreseh but frequently travelled from there to Pasuruan, staying for one or two nights each time in the warehouse and sharing a room there with some of the warehouse workers. The warehouse was supplied by a total of five vessels, Haji Matsahri's own prahu plus four Mandar prahus from Kota Baru on Pulau Laut, an island off the southeastern coast of South Kalimantan (Map 16).

Before the ethnic conflagration of 2001, Haji Matsahri's prahu, *Sumber Makmur* (Source of Prosperity), had been carrying timber from Pegatan, on the lower Katingan river (see Map 13, Chapter 4), to Pasuruan. When the Katingan became unsafe for Madurese vessels, with one Sepulu sailor murdered, Haji Matsahri decided to obtain his timber from Batulicin on the southeast coast of South Kalimantan (Map 16). This was not a port he had previously sailed to himself but he opted for it after discussions with the Mandar skippers who were also supplying his warehouse. In fact, Batulicin is a very convenient place for business, being regularly served by passenger ships from Surabaya. Despite not skippering his own vessel, Haji Matsahri was therefore still able to be directly involved in negotiations over his timber purchases. After being notified by telephone of the arrival of *Sumber Makmur* in Batulicin, he would travel there himself, leaving Sreseh at 4 a.m., in time to catch a modern fast ferry from Surabaya and arriving in Batulicin in the late afternoon of the same day. There he would stay on board on his prahu, living as one of the crew and not spending money on either accommodation or food stall meals. Although Batulicin is situated on an estuary, timber was delivered to the port by truck, there being sealed roads into the hinterland. For both crew and owner, being in port at Batulicin was much preferable to being moored, possibly for weeks at a time, alongside a sawmill in some remote place far from the amenities of a proper port town. Moreover, from the harbour at Batulicin it was only a short walk to the local branches of main banks.

The one drawback to obtaining timber from Batulicin was the high price of getting an SKSHH. Through any of the local shipping agents, the SKSHH cost Rp 120,000 per cubic metre, at least 20 per cent higher than through agents in most other places.[6] While the obvious remedy would have been for Haji Matsahri himself to conduct the transaction with the local forestry office, like most Madurese skippers he did not know the procedure. Moreover, he feared that his vessel might be physically checked if he tried to obtain the SKSHH

[6] Several skippers in Pasuruan complained to the author about the high cost of obtaining an SKSHH through agents in Batulicin. This included Mandar skippers from nearby Kota Baru, one of whom stated that he was fortunate to be paying only Rp 100,000 per cubic metre because his supplying sawmiller had arranged the permit.

himself. The latter prospect was especially worrying because the documentation for *Sumber Makmur*'s cargo was always for just 40 per cent of the actual amount carried.

Because the additional expense of obtaining an SKSHH at Batulicin could not be recovered on the highly competitive Pasuruan timber market, Haji Matsahri eventually turned for his timber to Marabahan on the Barito river, well upriver beyond Banjarmasin (see Map 13). However, although the timber was of good quality and there was no problem with excessive agent fees for organising the SKSHH, the supply at Marabahan was rather irregular. Unlike at Batulicin, most timber to Marabahan came down the river, towed by small narrow-beamed river craft (*kelotok*). During the dry season the water level gradually falls and it becomes increasingly difficult to get timber from the small tributaries close to where felling is carried out. Consequently, during the late dry season and the change of the monsoons prahus often had to wait for several weeks for the river to rise before they could obtain cargoes. Frustrated by this loss of productive time, Haji Matsahri decided to try to obtain timber from Kintap in the district of Tanah Laut on the southern coast of South Kalimantan (see Map 16). Neither he nor his skipper had been to Kintap, but since 2001 several Sreseh vessels had been obtaining good quality timber from there and all reports were favourable.

A problem with Kintap, however, was to find the place. Unlike most of the other timber ports of southern Kalimantan, the approaches to the Kintap river are almost featureless. There are no lights or beacons, no nearby islands, no distinguishing landmarks, no significant indentation in the coastline, and no township or large village visible from the sea. Moreover, although all the Sreseh vessels had compasses, none of the skippers owned a chart. A skipper wanting to go to a new destination would of course ask for directions from others who had made the voyage but for Kintap such directions could only be vague — "on the coast of South Kalimantan, about a third of the distance from Tanjung Selatan to Batulicin" — and would probably include advice to ask the way from local fishing craft. Indeed, Haji Matsahri's skipper did have to resort to this on his first voyage to Kintap.

On its last voyage for the year in November 2002, Haji Matsahri's prahu *Sumber Makmur*, with the author on board, motored out of the Blega estuary in the early morning then headed east. The early start enabled *Sumber Makmur* to clear the Madura Strait, including the busy

Kalianget area, before nightfall. If the vessel continued motoring at the same speed, however, it would reach the coast of South Kalimantan during the hours of darkness, which would be inconvenient and possibly dangerous. To avoid this, the skipper halted for a few hours near the island of Giliiyang, off the northeastern tip of Madura, anchoring offshore in fairly deep water. At 2 a.m. he started the engine and ordered the anchor to be hauled up, knowing from experience that by leaving Giliiyang at this time, his vessel would be close to, yet still at a safe distance from, the coast of South Kalimantan at dawn two days later.

During the first day out, *Sumber Makmur* passed to the east of the island of Masalembu (see Map 5, Chapter 1), the main 'signpost' on this route. As the darkness was fading on the second morning out from Giliiyang, the distant backdrop of the Meratus mountain range came into view and the skipper adjusted his heading towards the eastern extremity of this inland range. The prahu eventually closed with the low-lying coast, then headed along it at a distance of a kilometre or so. Fishing vessels, 12 to 15 metres in length, were numerous in the area. A red-and-white telecommunications tower came into view; although not shown on charts this was a useful sign along the way for the entrance to the Kintap river was just a few kilometres beyond this tower. A keen lookout was necessary because the entrance to the river is very narrow, only about 30 metres across. Moreover, the estuary veers to the right just a short distance in from the sandy coastal beach, so that from just a few hundred metres offshore the coastline at that point appears unbroken.

As well as being narrow, the entrance is shallow. It is no place for large vessels and many prahus have to wait for high tide in order to avoid grounding on the bar. On this occasion *Sumber Makmur* did ground heavily and it took much heaving on bamboo poles and the engine at full throttle before she managed to scrape through into deeper water. After passing the small fishing port of Muara Kintap, situated on the eastern bank about one kilometre from the entrance, *Sumber Makmur* turned left off the main river into an unlikely-looking tributary, narrow and densely lined with *nipa* palms. The journey continued for about three more hours along this winding waterway, with few signs of human activity except for the odd isolated house built on poles and the occasional fisherman in his canoe, tending his fish traps, until a series of small sawmills came into view. As their

Photo 33 Logs for processing at a sawmill on the Kintap river.

Photo 34 A tiny expatriate community: vessels from Sreseh rafted together at Kintap.

scrappy appearance indicated, these were unlicensed enterprises (the word in Indonesian to indicate such illegal status is *liar*, literally, 'wild'). The sawmills became more numerous, until eventually a small cluster of houses, hardly more than a hamlet, came into view on the west bank. This was the 'timber town' of Kintap. The settlement was more significant than it appeared from the river, however, as it was close to a main road and had a weekly market serving the population of the surrounding countryside. All of the houses were built entirely of wood, including wooden shingle roofs, and most of them were unpainted. For visiting Madurese mariners, used to the whitewashed brick and plaster houses of their home villages, these drab wooden dwellings were a reminder that in this sparsely settled land, virtually all human needs are obtained from the surrounding forest.

Although most of the 20 or so sawmills spaced out along both sides of the river were unlicensed, no one at Kintap seemed to think that there was anything remarkable about that. Most were owned by Banjarese entrepreneurs but four were Madurese-owned and a few others Bugis-owned. However, there was no ethnic preference on the part of prospective timber buyers. What mattered, from both sides of this resource frontier, was trust. Purchases were sometimes made on credit, with the debt to be settled when the vessel returned next time. If the vessel did not return and instead went elsewhere to load timber, the sawmiller had almost no chance of redress. Such incidents had happened. All the prahus loading timber at Kintap in 2002 were Madurese, mainly from Sreseh, with the remainder from Pasian, on the northeast coast of Madura, or Sepulu. As it happened, the sawmiller supplying the *Sumber Makmur*, a man named Hanan, was Madurese. Originally from Sapudi, he had migrated to Kintap as a young single man in 1978, coming by prahu to the nearby inlet of Sungai Danau. Despite never having met either Haji Matsahri or his son Marthawi, Hanan spoke of them as though they were friends of long standing, a reflection of the critical element of trust. Although he admitted that he would detest returning to Madura after so many years in Kalimantan, which he described as "the land of opportunity", he still clearly derived satisfaction from conducting business with fellow Madurese.

Hanan's first wife had been Banjarese, but after her death a few years ago he had married a Dayak woman. Because of the shortage of Madurese women, such inter-ethnic marriages were not uncommon in small places in South Kalimantan. Hanan's neighbour was also a

Madurese and he also had a Dayak wife.[7] Most unusually, however, in 2002 one prahu skipper from Sreseh married a Dayak woman in Kintap, even though he already had a wife in Madura. This individual, named Toha, had an unsavoury reputation in his home village because he was known to have married and permanently separated from several women, without divorcing any of them. He also had a background as a pimp and thug in Jakarta of which he made no secret — he had served time in prison and claimed to have once been featured on national television in the daily Jakarta 'criminal round-up'. However, he told me that he had moved away from his criminal ways. Instead, at the age of 38, he planned to become a timber trader. He had married a Dayak woman partly, he said, because he needed the comforts of home while away in Kintap but also because he wanted to win the trust of the local community so that he could get credit for an initial shipment of timber in his own right. There was no intention of cheating his potential supplier. Rather, for an aspiring trader with limited capital resources, the greatest hurdle was obtaining an initial shipment of timber. Once that shipment had been sold in Pasuruan and his debt to his supplier paid, Toha would be on his way as a successful timber trader. I liked Toha but I felt sorry for his new bride, who could hardly have been expected to have known of his history with women. As it happened, Toha's vessel was in Kintap when I was there and he invited me to his wife's place. I was mildly surprised to be taken to a comfortable two-storeyed house with a small food stall in front, very close to the river. However, when I was introduced to his wife, my surprise was much greater. Instead of the gullible girl I had imagined, I found a large Dayak matron with two adult daughters, a former widow who clearly had the measure of her man!

When timber first started to be exported out of Kintap, it all came downriver, towed by kelotok. In 2002 there were still plenty of kelotok towing logs at Kintap but the towed logs were of small diameter and intended for local consumption. Most of the wood

---

[7] When I was in Kintap in November 2002, there was a large sign at the rear of this neighbour's house, on the riverbank for anyone to see, which read as follows: "______ [Islamic name], of Sumenep, Madura, cannot live except with his wife ______ [Dayak name]." In the wake of the anti-Madurese violence in Central Kalimantan, this remarkable sign served as a reminder that in Kintap, at least, relations between locally resident Madurese and Dayaks were harmonious.

destined for Java was instead brought to Kintap by truck. Indeed, the main advantage of Kintap as a source for timber was its proximity to the main road running from Batulicin to Banjarmasin. The logs were mostly meranti, of good size and quality, from upland country about 50 kilometres inland. In Hanan's opinion, the timber industry could be sustained indefinitely if logs were taken selectively for timber extraction only, leaving the forest otherwise undisturbed. However, he also said that all of the timber he was dealing in was "waste", in the sense that the land where the logs were being taken from was being cleared for oil palm plantations. Although the supply was still plentiful, he was apprehensive about the future.

During the dry season the logs were carried out of the felling areas by ordinary two-wheel-drive trucks. This was not possible during the wet season, however, and at that time of the year buffalo teams were used to drag the logs to the nearest useable road, where they could be loaded onto trucks. The felled logs were cut into lengths of four to five metres because the trucks, with their short trays, could not carry logs longer than this. The moderately short lengths also simplified handling of the logs at the sawmill end, where there was no mechanised lifting equipment. However, in addition to short logs waiting to be milled, long pieces of ulin, about 12 to 15 metres by about 20 centimetres square, were regularly sold through Kintap. Cut directly from young trees, these were special items for which length was desirable. Larger diameter ulin logs were cut into short lengths for milling into planks.

Trucks could not reach most of the sawmills at Kintap because there was no serviceable road along the river banks. This was no problem, however, as the logs — excepting ulin — were simply dumped off the trucks alongside the river and rolled in, to be floated downriver to the owner's sawmill. Each sawmill had a ramp extending into the water and the logs were levered or winched up these ramps. Because the bandsaws of these small sawmills were not very large, the logs were first cut in half longitudinally then quarter-sawn, using chainsaws, before being milled into planks (Photo 35).

When *Sumber Makmur*'s skipper came to Hanan's sawmill in November 2002, it so happened that the local felling teams were then working in stands of keruing trees, so that this was the only timber available for sale. Marthawi had just taken a shipment of keruing a couple of months earlier and did not want any more at that time because it was not in strong demand. The problem with keruing was

Photo 35 Meranti logs quarter-sawn by chainsaw before being sawn into planks.

that it was rather expensive, yet with only moderate durability, unsuitable for exterior work in the tropical climate. Marthawi's customers were more interested in 'MC' ('emchay', meaning *meranti campur*, mixed meranti species), which although of about the same durability as keruing cost only about half as much. Keruing was a more consistent timber but meranti could serve just as well for most purposes. The logging gang offering keruing was lounging around in the otherwise empty Kintap market place, showing no interest in talk about 'MC'. The skipper and crew simply had to wait. Nevertheless, several days later Hanan managed to procure some meranti logs from another source and commenced sawing them into planks.

As the planks came off the saw, they were loaded directly onto the waiting vessel, moored alongside the sawmill. Whereas in larger timber supply ports the loading work was carried out by stevedoring labourers, in Kintap the crews had to load their vessels themselves. The crew members had the time available but the work was arduous and some were unhappy about having to do it. A particular problem for the crew of *Sumber Makmur* on this occasion was the loading of several long baulks of ulin intended for keels and endposts in vessel

Photo 36 Loading timber into the hold of a prahu at Kintap.

construction at Sreseh. Because of their length, these pieces could most conveniently have been stowed on the deck. However, because they were not to be declared, the skipper wanted them stowed in the hold, where they would be unlikely to be seen by officials.

Once the vessel was fully loaded, Hanan calculated the volume of timber taken on board and the price, including the cost of obtaining the SKSHH. Then, together with *Sumber Makmur*'s skipper, he telephoned Haji Matsahri, who in turn notified his shipping agent in Pasuruan of the amount to be transferred to Hanan's bank account. The initial payment was thus from the shipping agency's working fund, a common arrangement as the agency had the necessary capital available and the agency clerks had expertise in such matters. Haji Matsahri's shipping agent, Haji Solechin, was also Madurese and originally from Sreseh. He was one of the most successful agents in Pasuruan, although his business had declined in turnover since the peak year of 1999, when he owned five trucks to cope with the unloading from vessels using his agency. By 2002, three of his trucks had been sold to meet debts.

For vessels loading timber at Kintap, the SKSHH had to be obtained from the forestry office in the district administrative centre at Pleihari, several hours' drive away (Map 16). Hanan preferred to organise the permit himself. He did make a profit from this work but more importantly, he was known and trusted at the forestry office. If a prahu skipper were to try to arrange the permit himself, an official could become suspicious and want to verify that the amount carried was the same as that stated on the application form. With only 40 per cent of the actual load stated on *Sumber Makmur*'s SKSHH application, such a check would have been disastrous. Hanan also organised the port clearance, charging a fee for the work. On the way out of the river, timber carriers had to stop at the busy fishing port of Muara Kintap where there was a harbourmaster's office, and pay a fee before being allowed to carry on. According to Hanan, this fee was Rp 750,000 — with one-third going to the harbourmaster's office and the remainder being channelled equally to the local water police and the navy.

Upon completion of the voyage, Haji Matsahri received income from both freight and trade. First, as vessel owner, he profited from the freight charge of Rp 83,000 per cubic metre for the cargo of 120 cubic metres (round figures, in rupiah):

Gross freight earnings	10 million
Less operating expenses (fuel, victualling, agent fees)	6 million
Net freight earnings	4 million

The net freight amount was then divided into nine shares of Rp 444,400 each, one share for each of the four crew and two-and-a-half shares for both the vessel owner and the engine owner. As Haji Matsahri owned both vessel and engine, he received five shares, giving him a return of about Rp 2 million per voyage in 2002. This was not a true net profit, however, for he had to pay for engine maintenance and any spare parts needed.[8] The earnings of the crew were modest, at around Rp 440,000 per voyage, with a typical voyage lasting about a month. Nevertheless, such earnings were far better than would have been available from any other line of work in Sreseh. Additionally,

[8] Most Sreseh vessels used an eight-share division arrangement: four shares for the crew and two each for the vessel owner and engine owner. However, these vessels had four-cylinder engines. According to Haji Matsahri, he would have been unable to meet the costs of maintaining his six-cylinder engine on the 'small engine division'. The crew accepted this and enjoyed the benefit of faster passage times with the more powerful engine.

each crew member was allowed to carry one cubic metre of timber for his own purposes with no freight charge. Timber purchased by crew members was left on board until the vessel called in at Sreseh for a few days' rest before the next voyage. Timber is expensive in isolated areas and directly or indirectly, a profit could always be made on small lots like this. The skipper also received a bonus of Rp 275,000 for each completed voyage.

The freight charged by Haji Matsahri was very economical and significantly less than the charge made by Madurese vessel operators transporting timber to either Central Java or Sepulu. However, his greater profit came from trading (rupiah per cubic metre, at 2002 rates):

Purchase price (m^3)	315,000	
Freight charge	83,000	(paid to own vessel)
Port and agent fees (Kintap)	12,500	
Total outlay	410,500	
Sold on board	450,000	
Profit margin	39,500	

The above figures do not include the cost of the SKSHH, which Haji Matsahri disregarded in his own reckoning. The SKSHH cost was instead taken into account by his son Marthawi as manager of their timber warehouse, registered as a trading company (*usaha dagang*, UD). Haji Matsahri's prahu was not an asset of the company, being registered in his name only. This separation had the advantage that in the event of the vessel being apprehended for inadequately documented cargo, Haji Matsahri would appear as just the carrier. Liability would instead fall to Marthawi's business as the owner of the cargo, thereby protecting the critical asset of the vessel.[9] Nevertheless, in his arrangement with Marthawi, Haji Matsahri was the actual purchaser. The risk of the voyage was therefore his: if his vessel happened to be wrecked and the cargo lost, it would be he who would have to bear the loss. With a cargo of 120 cubic metres and a profit of Rp 39,500

[9] This was despite the fact that the consignee listed on the SKSHH was the shipping agency that had made the transferral of funds for the cargo on behalf of Marthawi's timber trading business. The agency would not be liable for shortcomings of timber documentation because such matters were beyond its control.

per cubic metre, the trading profit to Haji Matsahri in late 2002 was around Rp 4.74 million per voyage. This represented a return of 9.6 per cent on his capital outlay of about Rp 49 million, about the lower level of the returns for entrepreneurs bringing timber into Juwana. For a typical year of seven voyages, he could thus expect to make a return of about 70 per cent on his capital. Such a return seems quite lucrative but, as will be explained in Chapter 6, the risks were considerable.

On arrival into Haji Matsahri's warehouse, the timber entered the phase of circulation, taken care of by his son Marthawi. In addition to the purchase price of Rp 450,000 per cubic metre, Marthawi had to pay the SKSHH fee, amounting to Rp 4.5 million (Rp 90,000 per cubic metre, for 50 cubic metres out of the total of 120), plus stevedoring and trucking costs. The real cost of his shipment of mixed meranti in late 2002 was therefore over Rp 500,000 per cubic metre. If he sold it locally, he would not make a large profit. However, in Surabaya, where the timber transport regulations were strictly policed, the market price was higher. Pasuruan is only about one-and-a-half hour's drive from both Surabaya and the large inland city of Malang. By selling to these places, and other destinations in East and even Central Java, Marthawi was able to make a profit of up to Rp 100,000 per cubic metre after his port and trucking expenses — more than twice the profit margin obtainable from buying in Kalimantan and selling free on board in port in Java. With his warehouse being served by five vessels, bringing in a total of between 300 and 400 cubic metres of timber each month, the gross company profit would have been in excess of Rp 30 million per month. This was, of course, during the busy southeast monsoon; averaged over the whole year, the profit would have been nearer Rp 20 million per month. After deducting operational expenses, mainly rent for the premises and wages for five workers, the net profit was then shared equally between Haji Matsahri and Marthawi.

The big problem in earning such handsome profits through timber trading was the need for the correct documentation. When a load of timber arrived in port, the documentation for that shipment, stating not only the name of the buyer and the volume of the timber but also the number of pieces, sizes and species, had to be lodged with the port forestry office. This information would then entered into a 'register' for each warehouse. To move timber onwards by truck, a permit called the DPP (*Daftar Penggantian Pengangkutan*, list of

Photo 37 From truck to warehouse, Pasuruan. The unseasoned timber is mostly stored under cover.

change of transportation) had to be obtained from the forestry office, specifying the total amount, number of pieces, sizes and species for the onward shipment. This information would be duly noted in the register for the warehouse concerned. If the timber was fully documented in the first place, this was a minor administrative procedure; according to forestry officials in 2002, the cost of the DPP was only Rp 15,000 per truck (not per cubic metre).[10] But if the original documentation was inadequate, as was the case more often than not, the situation was complicated and risky.

The usual procedure in 2002 to deal with such a situation was to arrange a police 'escort' (*pengawal*) who would accompany the driver, conferring an apparent immunity to checks. At that time, the rate in

[10] Most merchants were actually paying much more than this for a truck transport permit, usually Rp 25,000 per cubic metre (not per truckload). However, this was a fee charged by the agency arranging the permit, not the official cost. Possibly a portion of this fee was given to officials so that the documents concerned would not attract close scrutiny.

Pasuruan for an 'escort' was Rp 150,000 per truckload — easy extra money for the police and cheap insurance for the merchant. Some merchants chanced deliveries without an escort but the risks were high. In September 2002, a Madurese timber merchant named Baidawi was arrested because a truck shipment from Pasuruan to Malang was found to have no documentation. His premises were subsequently checked and timber worth around Rp 100 million was seized. He was reportedly released after one month in prison, only because his wife managed to make a cash payment of Rp 20 million to the police. The informant concerned, another Madurese timber merchant, stated that his own shipments out of Pasuruan for the previous few years had always been accompanied by police escorts. His wife, daughter of a well-known Sreseh family and a relative of the unfortunate Baidawi, added that she had no sympathy for the latter because he had been "too miserly to pay 150,000 rupiah".

Marthawi also had his share of problems with truck transport. In September 2002 a truck carrying timber from his warehouse was detained at a road block in Central Java. The driver had a DPP permit with him but the document did not, as required, state the species of timber. Knowing that such an omission was most unlikely for a DPP arranged through the proper channels, the police telephoned Marthawi, who had to pay Rp 6 million before the truck was allowed to go on its way. According to Marthawi, the load was legal but the driver, a private contractor, had falsified the permit to show five cubic metres less than the actual amount, planning to steal the difference. Unfortunately for him, his temerity was not matched by an appropriate skill as a forger. Marthawi did not complain to the police about the driver because he did not want to invite unwelcome scrutiny but the driver was not contracted again. Marthawi was particularly vexed over this matter, not just because of the driver's dishonesty and the loss of a large sum of money but also because, as he claimed, the police were acting on a tip-off from a rival timber merchant in Pasuruan.

Such underhand business tactics have indeed been a feature of the intensely competitive Pasuruan timber market since the crackdown on the traffic in illegal timber. This was illustrated by the case of a Bugis timber merchant, Haji Aliansyah. This individual, a former prahu sailor from Pegatan in South Kalimantan, had saved up money until he was able to set up his own timber trading business, P.T. Bintang Cemerlang, in Pasuruan in the early 1990s. Despite his modest start, his business approach was modern, in marked contrast to the other timber merchants around him. While most of his rivals had only an elementary education, Haji Aliansyah commenced business management studies at university and in 2001, at the age of 41, obtained his bachelor's degree. This in itself was a remarkable achievement, given the rarity of mature-age students in Indonesia. In further contrast to other timber wholesalers in Pasuruan, Haji Aliansyah not only employed five women to run the clerical and sales side of his business but also encouraged and assisted them to obtain tertiary qualifications. As his business grew, he also invested in vessels and engines, ensuring a steady and plentiful supply of timber. In 2001 his business was supplied by up to 17 vessels, some of which he owned outright while in some other cases he owned the engines. He also owned a large bandsaw and a forklift, items not possessed by most other timber firms in Pasuruan. Much of the timber coming into his warehouse was sent on to Jakarta, where the price was highest. Unlike his rivals, he never had problems with checks on his shipments because he cultivated good relations with senior police and forestry personnel. It seemed he was untouchable. In the Madurese-dominated Pasuruan timber trade, resentment smouldered.

An opportunity presented itself in 2001 during Ramadhan, the fasting month, after the appointment of a new local police chief who was apparently determined to root out the many timber merchants who were defrauding the state. Haji Aliansyah had not yet established a friendship with this person, perhaps deeming it more polite to wait until after Lebaran, the festive period at the end of the fasting month. At the time, Haji Aliansyah had been bringing in some baulk timber and was milling this with his bandsaw into smaller sizes. Another timber merchant of his acquaintance, a Madurese, asked Haji Aliansyah to process a few baulks for him because he himself had no sawmilling equipment. Haji Aliansyah agreed and the baulks were duly delivered, without documentation. The following day there was a police raid and naturally enough, the lack of documentation for this timber was discovered. The premises were sealed forthwith and Haji Aliansyah was arrested. He subsequently served five months in prison. After being released he moved to East Kalimantan, where he set up four sawmills. A year after his arrest his warehouse in Pasuruan was still not back in business, although one senior staffer had been retained as office manager, along with a couple of yardmen. The forklift had not been sold because Haji Alianysah hoped to eventually reopen the premises. According to both the office manager of P.T. Bintang

Cemerlang and officials from the harbourmaster's office at Pasuruan, the downfall of Haji Aliansyah was the result of a trap set by some of his rivals. He had become too successful, using business acumen and skills that they did not possess. Such was the nature of the timber business in Pasuruan after the police crackdown in 2002, a cut-throat trade fraught with rivalry and punctuated by personal business calamities.

Sungai Kaki to Sepulu: The Limits of the Resource

After the imposition of high district levies on timber in 2000 as part of local-regional efforts to generate revenue for the administrative district of Kotawaringan Timur (see Chapter 4), the timber entrepreneurs of Sepulu turned for their timber supplies to the Kaki, a small river to the east of the Katingan in the southeast corner of Kotawaringin Timur (Map 17). Because there was no township on the Kaki, the sawmill area there was referred to by both prahu sailors and locals as Sungai Kaki (Kaki River). Some Sepulu prahus had already been taking timber from Sungai Kaki for several years, as well as from places along the Katingan. After 2000, however, Sungai Kaki became the main source for timber landed at Sepulu because its remoteness made it easier to avoid paying the district timber levy. Set in a sparsely populated peat swamp area with no road access and no service by coastal passenger boats, Sungai Kaki was a backwater rarely visited by officials.[11]

Although the peat swamp forest between the Sebangau and lower Katingan waterways was designated as production forest, it had held little interest for large-scale logging concerns because of the generally indifferent quality of the stems. Nevertheless, after 1990 the area became subject to intensive illegal logging (McCarthy, 2001b: 5). The sawmills at Sungai Kaki, all unlicensed, were owned by Banjarese entrepreneurs. From the outset the main carriers from this isolated waterway were prahus from Sepulu. Some vessels from Sreseh had

[11] Regular passenger services around the southern coast of Kalimantan are provided by so-called 'speedboats', fibreglass boats with large outboard engines and seating for about 20 passengers. While such boats travel between Katingan Kuala and Kuala Kapuas, to the west and east of Sungai Kaki, they do not put into the latter because of insufficient demand. To do so would also involve a considerable deviation from the Katingan–Kapuas service.

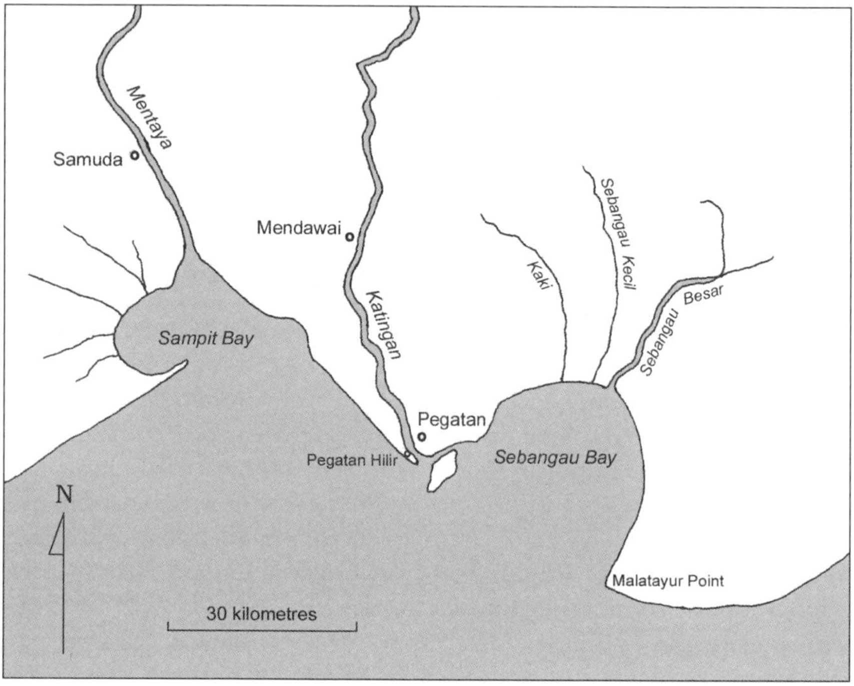

Map 17 Sampit Bay to Malatayur Point

taken timber from there during the 1990s but eventually they all turned for their supplies to South Kalimantan (Marabahan, Kintap or Batulicin), where the timber was of better quality. For the Sepulu timber entrepreneurs, however, what mattered most was that timber from Sungai Kaki was cheap. The more cheaply the timber landed in Sepulu, the more easily it could be sold to buyers from Surabaya and other parts of East Java. The indifferent quality of the timber mattered little because lower-quality timber has its own market niche, especially for use as concrete formwork on construction sites.

As well as the advantage of avoiding paying the regional timber levy introduced in 2000, Sungai Kaki was, by virtue of its remoteness, the only place in Central Kalimantan at which *golekan* vessels, with their unmistakable Madurese styling, could take on timber in the wake of the ethnic conflict in 2001. Some Sepulu vessels had also been obtaining timber from the Sebangau river, to the east of Sungai Kaki and located in the kabupaten of Kapuas. In 2001, however, timber from the Sebangau became at least as expensive as that from the Katingan after the Kapuas administration followed the example of Kotawaringan Timur and imposed its own timber levy (McCarthy, 2001b: 9). Since then the Sebangau has been prohibitively expensive for small craft, although a few large Sepulu vessels carrying around 300 cubic metres continued to obtain timber from that river, their larger capacities enabling the higher cost to be absorbed. A few Sepulu entrepreneurs turned to Kintap and Asam-asam (Map 16) in South Kalimantan but these places were further away from Sepulu and the timber was not particularly cheap. For most of the Sepulu fleet, Sungai Kaki remained the best option.

The land around Sebangau Bay is all low-lying, so that the entrance to Sungai Kaki, which is just as narrow as that of the Kintap river, cannot be seen until one is close by. Nor is there any fishing fleet at Sungai Kaki to help point out the way. To get there, skippers of Sepulu prahus headed first for the broad low-lying promontory of Malatayur Point, and thereafter followed the shore of Sebangau Bay until they reached Sungai Kaki. From Sepulu to Malatayur Point took about 30 hours with a good engine. The prahus used to leave Sepulu in the morning, with no great haste, and anchor in the afternoon of the following day off Malatayur. This allowed them plenty of time the next morning to find the entrance to the river and get settled in near the sawmill area before nightfall. Unfortunately, local pirates, probably opportunists rather than professional maritime bandits, came to take advantage of this routine. Many Sepulu vessels were boarded during the night while anchored off Malatayur Point and their crews robbed. The crews soon armed themselves with Molotov cocktails and some informants claimed to have used these successfully to ward off pirate 'speedboats'. A better solution, however, was to plan the voyage so that it would not be necessary to anchor off Malatayur Point, reaching there instead in the early morning and continuing straight on to Sungai Kaki. After the mid-1990s this became standard practice and vessels left Sepulu at night rather than during the day as previously. The pirates were reportedly flushed out by the local authorities but they may have simply desisted from their activities after prahus had ceased anchoring off the point for the night.

The sawmill area at Sungai Kaki was situated only about one kilometre from the entrance, where the water was still brackish and not pleasant for washing oneself. The settlement, such as it was, was singularly uninviting. There was no market, no fields, no road, just a sawmill camp with large open sheds on both sides of the narrow river, a line of purpose-built workers' huts on the east bank and the ground all around covered with sawdust and mud. There were few other houses apart from workers' huts. During my visit, early in the wet season, a heavy overcast sky and intermittent rain added to the dreariness of the scene. As night closed in, along with steady rain, the prahu crews stayed on their vessels. When a friend, a hardened middle-aged mariner with experience of Sungai Kaki, heard that I intended to go there on a prahu from Sepulu, he warned me: "Sleep on board the prahu, so you will be among friends." In the event I accepted the offer of the local *kampung* head to stay in his house, but I came to understand what my friend had meant. It was not that the local people were in anyway hostile to Madurese sailors; indeed, they depended upon the Madurese vessels that called there. Yet for the outsider there was no comforting sense of entering a true community and no social resemblance to Kintap, for example, which, despite its shortcomings for visiting prahu sailors, was in no way threatening. In Sungai Kaki there were no Madurese residents and interaction between the sailors and sawmill workers was minimal. While it had probably never been pleasant to visit, like many other remote places where people stay only for work, in early 2003 there was a distinct air of despondency about the place. A woman whom I spoke with told me that she had been recruited for work in Sungai Kaki from Banjarmasin a few years earlier. She had not intended to stay for long but having married a local Dayak man, she felt trapped there and could not imagine how she would ever get away from the place. A man, a sawmill worker, asked me directly if it was true that in Australia prisoners are given clothing, food and health care. When I replied that it was true, he said, without a flicker of a smile, "In that case, that [prison in Australia] would be better than staying here." Such comments reflected the mood of the place. But what most darkened the mood was that the resource upon which everyone depended was drying up. Most of the sawmill workers were then idle because of a shortage of logs. The few logs that were arriving were small and of poor quality, producing reject-grade planks suitable only for temporary work. Later on, as the river levels rose during the wet season, more logs would become available. Nevertheless, the prevailing view at Sungai Kaki was that the local forest could offer a livelihood for only another two years. Logs brought to the sawmills at Sungai Kaki came only from the immediate environs of the Kaki river itself, which is only a small waterway, and the surrounding forest had already been heavily logged.

There were no roads. The situation was thus quite different from that at Kintap which, although also situated on a small river system, was connected through a network of roads to an extensive upland forest area.

The prospect of the sawmills closing down was of course far more serious for the local people than for the Madurese sailors. The latter nevertheless had their own reasons for discontent. Because the wet season had just started, the river level was still low, resulting in few logs reaching the sawmill camp and a long wait to take on a full cargo. The success of the Sepulu vessel operators over the previous decade had to a large extent depended on quick turnarounds, with four trips a month the norm. The fast turnarounds had allowed the owners to pay their crews in wages rather than a profit-sharing arrangement as applied elsewhere, which in turn enabled the timber to be landed more cheaply than would otherwise have been the case. By these means, as well as by avoiding the expense of obtaining full documentation, the Sepulu timber traders had managed to undercut the Surabaya market. In 2002–03 the wage rate for crew in Sepulu was Rp 150,000 per man per voyage, with as much as Rp 400,000 for the skipper. While Rp 150,000 per voyage was very low compared with the earnings of crew members elsewhere, being able to make four voyages a month had compensated for this. However, because of the limited supply of logs arriving in the sawmill camp in November and December 2002, it was taking a whole month for just one voyage. Consequently, earnings of crew members were barely adequate for survival. In April, after a three-month lay-off for the Sepulu vessels, there would once again be a 'timber flood' coming down the swollen Sungai Kaki. Nevertheless, everyone knew that four voyages a month would never happen again.

In addition to the problem of supply, the economic viability of the timber trade from Sungai Kaki was now threatened by the vigilance of the officials in the recently created administrative district of Katingan. In the early days of the timber boom from Sungai Kaki, vessels sailed just as their skippers pleased, without any administrative protocol. Following the crackdown on the lack of transport permits and state revenue evasion, skippers had been obliged to obtain an SKSHH, but until 2002 such documentation had been for only 30 per cent of the cargo, the remainder being illegal. The document was arranged by the supplying sawmiller, there being no shipping agents in Sungai Kaki. According to the Sepulu harbourmaster, however, the

Photo 38 A large motorised *golekan* at Sungai Kaki.

SKSHH documents for cargoes arriving in Sepulu had been issued in Banjarmasin, which of course is not in Central Kalimantan. From Sungai Kaki it is a long way to Banjarmasin, but the local sawmillers owned or had access to small fast outboard boats and they also knew the procedure and officials in Banjarmasin, which was really home territory for them. More importantly, unlike in Kotawaringin Timur, there was no special district timber levy payable in Banjarmasin.

Although some officials in Kotawaringan Timur must have been aware of this loss of revenue to their district, Sungai Kaki was such a difficult place to reach that nothing was done about it until late 2002, after the establishment of the new administrative district of Katingan (see Map 14). Although an SKSHH from Banjarmasin was perfectly legal in Java, the officials of Katingan had strong reasons for wanting the permit to be obtained in their own administrative district. First, the district timber levy of Rp 50,000 per cubic metre could then be enforced because an SKSHH would not be issued without this levy having been paid. Second, although all the state timber taxes (PSDH, DR) had to be remitted to Jakarta, 32 per cent of the amount remitted would subsequently be given back to the source district. To ensure the collection of revenues from timber leaving the district, 'combined posts' staffed by forestry, police and harbour officials were set up at strategic points. Whereas clearance to sail used to be given by a harbourmaster — the nearest one to Sungai Kaki being stationed at Pegatan Hilir — clearance now had to be obtained from the 'combined post' and would not be issued unless an SKSHH could be produced from Katingan.

If the timber coming from Sungai Kaki had been of high quality, the entrepreneurs of Sepulu might have been able to cope with the additional cost of the district timber tax. Instead, the unanimous view in both Sungai Kaki and Sepulu was that the district tax was unreasonable for such low-grade timber and left an unviably low profit margin on the Java market. This view was understandable because the district timber levy was set at a fixed level, irrespective of both the species and the quality of the timber. Nor did the state timber taxes take the quality of timber into account. It was therefore not surprising that, despite the efforts of the Katingan officials to close the Banjarmasin loophole, the Sungai Kaki sawmillers continued to obtain their documentation in Banjarmasin. According to the Sepulu harbourmaster, clearance documents for vessels departing Sepulu for Sungai Kaki always stated the intended destination as a place called Kerengbangkirai, an administrative subdistrict to the south of Palangkaraya, rather than the much closer Pegatan Hilir, where the officials were regarded as 'too strict'. On their return to Sepulu, however, instead of a clearance to sail from Kerengbangkirai as might have been expected, these vessels routinely carried a port clearance, as well as an SKSHH, from Banjarmasin. Extraordinary as this was, from the Sepulu harbourmaster's position there was nothing illegal about it. Indeed, he was sympathetic to the argument that the district timber tax being levied in Katingan was inappropriate for low-grade timber. Nevertheless, by mid-2003 he was less than impressed by the level of compliance in Sepulu with the national timber transport regulations. Throughout 2002 the level of compliance had been consistently high at 90 per cent of the actual volume of timber carried. However, as 'operations' in Sepulu by police and forestry teams had become less frequent, the level of compliance with the regulations had declined, so that by mid-2003 most prahus were again carrying documentation for only around one third of the actual amount carried.

If the authorities concerned had presumed that 'socialisation' of the timber transport regulations had been achieved in Sepulu after a year of virtually complete compliance, they were clearly mistaken. Just as the timber traders of Sepulu had little choice but to break the law in Katingan by getting their transport permits elsewhere, so too their reversion to massive understatement of cargoes in order to minimise payment of state taxes was understandable. The problem was that so many merchants in Pasuruan were still obtaining their timber cheaply by understating the volume of shipments to that port. Under such circumstances, the entrepreneurs of Sepulu could not compete on the Surabaya and Malang markets unless they followed the same strategy. As Haji Rofi'i, head of the desa of Prancak in Sepulu said, "It's like beer. Beer is supposed to be banned in many places on Madura, yet it still comes in. In the same way, illegal timber keeps coming in through Pasuruan, undermining our business." Yet he did not blame the Pasuruan merchants so much as the East Java provincial government and was bitter about what he regarded as its overly zealous attitude on timber imports, unlike in Central Java, where he averred the timber trade was "completely free".

Although the timber merchants at Sepulu were evidently finding their own way around these economic problems, the situation was more difficult for vessel operators without any trading interest. One of the many vessel owners of Sepulu in this category was Haji Saiton, whose *golekan* was in Sungai Kaki in December 2002. Lacking sufficient capital to set up as a trader, his income was from freight alone, with his 'timber boss' a man from Klampis, a short distance to the west of Sepulu. However, Haji Saiton did own the vessel's engine himself. His house was very comfortable and well furnished, suggesting that his shipping business had been successful. Yet he denied this, saying that his house had been financed not from shipping profits but from a long stint working in Malaysia. In a corner of the lounge was a ship-to-shore radio, with another on board his prahu. The radio sets had cost him Rp 2 million for the pair, second-hand. While the safety regulations for prahu shipping stipulated that all motorised prahus over 35 gross tons (about 100 cubic metres) must have two-way radios, Haji Saiton had not bought the radios for reasons of safety at sea. Instead, he had bought them so that he could warn his vessel to stay at sea if an 'operation' was underway at Sepulu or, alternatively, that it was safe to come into port. This was normal at Sepulu, with most of the vessels there by then carrying radios as essential 'economic equipment'.

In mid-2002, Haji Saiton's prahu had been intercepted at a 'post' on the Sebangau river, where good timber was still plentiful. Haji Saiton claimed that although an SKSHH had been obtained for the cargo, a special district levy (*taktis*) of Rp 9 million was demanded before the vessel could proceed. This payment became the responsibility of the timber trader from Klampis, who thereafter decided that it would be better to obtain his timber from Sungai Kaki. As vessel owner, it made no difference to Haji Saiton whether his prahu carried timber from the Sebangau or Sungai Kaki because the freight charge was the same for both, Rp 90,000 per cubic metre. This was the standard freight rate for Sepulu vessels. However, with his vessel having a capacity of only 100 cubic metres, in 2002–03 that freight rate was no longer sufficient to give Haji Saiton an acceptable return, as the following account shows (round figures, in rupiah):

Gross freight earnings	9 million (100 m^3 × 90,000)
Less expenses:	
Fuel	5 million (2,300 litres × Rp 2,200)
Victualling, port costs, etc	1.5 million
Wages (ordinary crew)	0.75 million (150,000 × 5 crew)[12]
Skipper's wages and bonus	0.4 million
Net freight earnings	1.35 million

The above figures would hardly encourage a prudent investor to become involved in the shipping business. While the net earnings represented a return of 17.6 per cent on the outlay (Rp 7.65 million) over a period of one month, this figure did not take into account the costs of engine maintenance and vessel depreciation. Returns had previously been much better than this.

Two sets of factors had contributed to the decline in profitability. The first concerned the cost of operating a motor vessel. Motorised prahus had given excellent returns until 1997, but after the rupiah plummeted the cost of engines and spare parts, all of which had to be imported, rose sharply. Only diesel fuel remained cheap at well below world parity prices. Eventually, however, the price of diesel did increase substantially as Indonesia came under pressure from the International Monetary Fund to phase out domestic fuel subsidies.

[12] A complement of six (five ordinary crew members plus the skipper) was typical for Sepulu. In Sreseh most vessels of similar size had a crew of only four including the skipper.

Consequently, in late 2002 it was costing Rp 440,000 on Madura for a 200-litre drum of diesel, approximately twice what it had cost two years earlier. This would not have mattered if the selling price of timber in Pasuruan had increased accordingly. However, because of the extremely competitive nature of the timber market, coupled with the generally weak demand in 2002, no timber merchant was prepared to increase his prices over those of anyone else. Instead, the extra costs simply had to be absorbed, severely reducing voyage profits.

The second set of factors affecting freight profitability to Sepulu concerned the forest resource itself. The modest voyage profit shown above might have been acceptable if four voyages were still being carried out per month, as had been common during the 1990s. Due to the serious decline in the availability of logs at Sungai Kaki, however, it was now taking a whole month to obtain a single load.

Crew members were similarly affected by these conditions. As explained above, Rp 150,000 for a month's work was very low by comparison with what prahu crews elsewhere were receiving. While a change to the profit-sharing arrangements as used in other parts of Madura would have improved crew earnings, such a change would have driven many vessel owners at Sepulu out of the shipping business altogether. Haji Saiton was extremely pessimistic about the future. He said that he wanted to sell his vessel, adding gloomily, "If you know of anyone who wants to buy a prahu, mine is going cheap." Like many prahu operators, in Sepulu and elsewhere, he had come to realise that in the rapidly changing circumstances of modern Indonesia, vessels that would once have been regarded as large were now too small to be economically viable without a trading interest as well.

Conclusion: Islam and Commerce in Madura

Because of the dense population and the limited suitability of the land for food crops, the Madurese have for centuries been an outward and trade-oriented people. Mansurnoor, an authority on Islam in Madura, suggests that the pattern of Madurese involvement in interinsular trade and migration was already established before the advent of Islam in Madura (Mansurnoor, 1990: 8). Nevertheless, Islam undoubtedly enhanced the commercial character of the Madurese. According to de Jonge, who in the late 1970s carried out an economic anthropological study in a town on the south coast of Sumenep, religion had served as a catalyst for commercial involvement in Madura. In the area where he was based, trade and commerce expanded significantly during the last quarter of the nineteenth century, at the same time as religious consciousness deepened and many locals undertook the pilgrimage to Mecca (de Jonge, 1989: 277). Castles (1967) and Geertz (1963) both note a similar connection between orthodox Islam and commercial entrepreneurship in nearby Java.

Whereas in the case of Christianity the Protestant Reformation marked a departure from earlier disapproving attitudes towards commercial enterprise, in Islam trading has always been regarded as an honourable livelihood, perhaps because the prophet Mohammad had himself been a trader in his younger days. Moreover, in Southeast Asia the spread of Islam, under the auspices of the Sultanate of Malacca, paralleled the expansion of trade in the region (Thomasz, 1993: 81–9). During the early period of Islamisation of the archipelago, the role of religious preachers was probably limited mainly to the conversion of local rulers, while in the port towns Muslim traders were already influencing others around them to adopt Islamic practices (Reid, 1993: 159). Probably some early preachers were also traders, and at least one famous preacher in Madura in the late fifteenth or early sixteenth century came from a trading background (Mansurnoor, 1990: 21).

Given this historical association between Islamisation and commerce, it is hardly surprising that in Madura there is no conflict between religion and striving for commercial profit. Indeed, kiai are frequently requested to give their blessings to commercial ventures, including prahus. Nevertheless, kiai themselves are not involved in commercial enterprise, for such involvement would be regarded as inappropriate for their status as religious elites. Nevertheless, this does not mean that they are indifferent to commerce. On the contrary, they depend to a large extent on donations from successful businessmen for the construction and maintenance of their residential schools (*pesantren*), as well as mosques and prayer houses (de Jonge, 1989: 277).

It is possible that in earlier times some kiai may have even been involved in maritime commerce, although they would probably not have gone to sea. Evidence for this comes from an old *golekan* at Telaga Biru which was in fact owned by a kiai. This traditional vessel, the only remaining example of the large sailing fleet which existed on the northwest coast until the early 1980s, stands under cover on display at Telaga Biru. According to informants at Telaga Biru, including a retired desa head, this prahu was owned by the late Kiai Syaichona Cholil, who formerly lived in Telaga Biru but later moved to the town of Bangkalan. His son, Kiai Abdullah Sahal, was in 2002 a senior kiai in Bangkalan. The vessel was named *Sarimunah*, a woman's name, after the owner's wife. Although completely traditional, as if fashioned by the same hand that had built all the other sailing *golekan* at Telaga Biru, *Sarimunah* was reportedly constructed of only the best-quality teak, which is perhaps why the vessel survived. It was claimed that when Kiai Cholil announced that he would have a prahu built for himself, people from the hinterland came forth with so much teak, given freely for the project, that there was enough to build a local mosque as well. Locals at Telaga Biru like to tell stories about this prahu and its supposed magical properties, such as when a man who stole the thick teak planks of the loose deck in the hold became afflicted by a mysterious malady, which abated only after he returned the planks to the vessel. While these tales are apocryphal, *Sarimunah* stands as a testament not only to the maritime culture of the people of Telaga Biru but also to the ongoing links between Islam and maritime commerce.

6

Entrepreneurship and Risk

Because of the dearth of ethnographic studies of groups involved in prahu shipping, as noted in Chapter 1, the representations in the literature of this unique maritime institution have amounted to little more than a series of glimpses from the outside, casting almost no light on the motivations, strategies and experiences of the men who actually operate the vessels. This chapter addresses this shortcoming with respect to the Madurese with profiles of six individuals interviewed during 2002–03, sketching their careers and in the process providing insights into such aspects as the significance of family background, business philosophy and approach, and the financial and physical risks involved in their maritime vocation.

The six cases presented here are not necessarily typical of the maritime Madurese. They have been chosen partly on the basis of their importance, with three of the persons featured having significantly influenced the course of events around them, yet also for their difference from one another. Each of the main trading prahu centres of Madura — the northwest coast, the southwest coast and the eastern islands — is represented. The names of some of these persons have already been mentioned but with no elaboration about them or the course of their careers. Following the profiles and drawing from them, the latter part of the chapter presents some observations on Madurese maritime entrepreneurship with emphasis on the economic uncertainty of vessel ownership.

Pak Yakob

Pak Yakob is a former head of the desa of Telaga Biru. He held the position for 26 years, retiring in 2000 in favour of a nephew. Nevertheless, he remained popular and locally influential and his personal backing was essential for his nephew to succeed him as desa head. In 2002, in his mid-sixties, he retained a commanding presence. With a forthright manner of speaking, robust physique and a patch over one eye, he was still an example of the tough (*blater*) type of local leader, the exception to the pattern of local leadership in Madura. And indeed, Pak Yakob's career has been more remarkable than that of the typical local-level administrative head.

He was born in Telaga Biru, the son of a prahu sailor. Like most other young men of his village, his education was elementary and in 1955 he went to sea at an early age, crewing on a small *golekan* shipping cattle to Pontianak in West Kalimantan. As usual in seafaring families, he did not work on the same vessel as his father. Going to sea was more than simply to earn income — it was also an opportunity to experience the wider world. In such a small, egalitarian and closely confined community, sons might feel uncomfortable working alongside their fathers.

In those days no Telaga Biru vessel had a compass or navigational chart. The main navigational 'signpost' for the outward journey from Madura was the island of Karimata: when Karimata came into sight, vessels bound for Sumatra would head to the left (west), while those bound for West Kalimantan would turn to the right (north). As well as such key landmarks, skippers relied on the appearance of the water to judge their position. Although the skipper was sometimes a young man, given the position because he was a close relative of the owner and therefore likely to be trusted, there was a lot of sailing experience in each crew and vessels rarely went astray. Nevertheless, Pak Yakob recalled that during his first year at sea, the prahu on which his father was working inadvertently sailed to Johore on the east side of the Malay peninsula rather than to South Sumatra as intended.

The first vessel Pak Yakob worked on carried ten to fifteen head of cattle, tethered on a removable deck in the hold. Larger *golekan* could carry more than twice that number of cattle but in those days small prahus predominated. For ventilation, a large hatch was usually left open on either side of the steeply gabled deckhouse roof. Some salt, brought by truck from Sumenep, was carried below in the bilge, while the crew also brought batik from Telaga Biru for sale in Kalimantan. Large amounts of water and grass had to be carried to maintain the animals in good condition for the duration of the voyage, which was a week to ten days. Feeding the cattle was a big chore, keeping the crew occupied intermittently throughout the day. In the closely confined conditions and with the highly strung nature of Madurese cattle, moving around among the animals while at sea required care. Especially was this so when giving out water, for the cattle would get very thirsty during the voyage and sometimes jerk violently in their haste to get at the water. Pak Yakob can attest to the danger of this work because it was while tending cattle that he lost his left eye, gored by a horn.

After the cattle, salt and batik had been sold in West Kalimantan, the *golekan* would set out empty on the return voyage. A round trip would take between 30 and 40 days, so that 6 or more voyages were possible before the onset of the northwest monsoon. In November the Telaga Biru prahus would all be hauled up on to the beach, where they would remain until March. Because the southeast wind was directly against them on the homeward leg from West Kalimantan, the sailors would in the first instance head south to Java, after which they could take advantage of the diurnal coastal effect upon the prevailing wind to enable them to work their way eastward along the coast until they reached Madura. They would break the homeward voyage in Java, usually in either Jepara or Pekalongan to buy rice, but also for rest and recreation. For the latter purposes, Pak Yakob recalled that 'komplek' — brothels — were popular with the sailors during these visits, probably the more so because such institutions were (and still are) virtually non-existent in the strict social climate of Madura. Meanwhile, back in Madura, the womenfolk passed their spare time making batik "so they wouldn't miss their husbands too much", as Pak Yakob put it. As a result of these frequent visits to Jepara and Pekalongan, close social ties developed between those two places and Telaga Biru. Pak Yakob's grandfather, also a sailor, had married a woman from Pekalongan and settled there before the Japanese Occupation.

Until the late 1970s, Telaga Biru had scarcely been touched by the social and economic changes taking place elsewhere in Indonesia. The character of the place was still very traditional, as exemplified by the large fleet of engineless *golekan* and a fleet of much smaller but similarly traditionally styled vessels for fishing. Prahu shipping and fishing provided adequate livelihoods for the local population. With a fleet of about 150 to 200 sailing trading vessels, all requiring crews of 8 to 10 men, there were ample employment opportunities for able-bodied men. There was also plenty of work in the construction and repair of vessels. Life was materially simple but generally satisfying.

Nevertheless, there were risks involved in seafaring. Indeed, on one dire occasion in 1963, Pak Yakob himself thought that his life was about to be cut short. The *golekan* that he was crewing on was carrying 38 cattle, bound for Pontianak. A couple of days out from Telaga Biru, the wind blew during the night with great force so that the prahu could only run before it, the mainsail furled and the small foresail set to enable some degree of control. Just at dawn a huge crest crashed down over the stern and blasted through the after doorway and into the hold, half-flooding the vessel in an instant. Desperately the crew tried what would otherwise have been unthinkable — jettisoning their valuable cargo, the cattle. However, in the raging sea they were unable to get any of the terrified animals out of the flooded hold. Other waves washed across the stricken vessel, until it went down at about 7 a.m., with all 38 cattle still tethered inside. In the last frantic moments, someone had managed to untie the *sekoci*, the small canoe-like craft that served as a tender, from its seagoing position on the deckhouse roof. As the prahu foundered, Pak Yakob found himself in the water, clinging to the upturned *sekoci*. He was soon joined by two others, one of whom was his younger brother. There was no sign of the other five crew members.

Somehow the three men managed to hang on to the tiny flat-bottomed, double-ended craft, just four metres long by one metre wide. After it had been bought in Pontianak, an extra plank had been added to increase its capacity, in the event a most fortunate modification. As the day wore on the seas moderated to the extent that the men could bail the *sekoci* with their hands and flip themselves into it. They were alive but their situation was precarious. They had no food or fresh water, no sail or paddle, there was no land in sight, and the prevailing wind would take them further out to sea. After drifting helplessly in heavy seas for two days and nights, bailing frequently, the man who was not Pak Yakob's brother suddenly pitched himself into the sea and drowned. For two more days, Pak Yakob and his brother drifted. Then, on the fifth day, they saw a vessel in the distance, a *golekan*. It was heading away from them but soon afterward the southeast wind died away and a gentle breeze set in from the north, induced by the warmer air rising over Java to the south. The *golekan*, which as it happened was from Telaga Biru and had left Jepara that morning, was forced to tack, and in the light conditions was carried by the sea breeze directly to the little *sekoci*.

After their miraculous rescue and return to Telaga Biru, Pak Yakob's brother was reluctant to go back to sea and it was a month before he joined another vessel. Pak Yakob, however, was phlegmatic about his ordeal and joined another prahu within a week of his return. "I wasn't afraid", he said, "because I knew it was God who had saved me... God did not want me to die yet."

Perhaps it was this sense of being marked out among his peers that led Pak Yakob to stand as a candidate in the election for desa head in 1974. Certainly, his role as desa head was to be distinguished by an uncommon economic leadership. At that time, a few large vessels of non-traditional type were being built at Telaga Biru for carrying timber from Kalimantan to Surabaya. Pak Yakob reasoned that instead of prahus from Telaga Biru carrying timber to Surabaya, the timber could be discharged and sold at Telaga Biru, thereafter to be transported by road and the vehicular ferry from Kamal to Java. He encouraged local entrepreneurs to set up as timber merchants, allowing them to build warehouses along the port foreshore. The timber business at Telaga Biru was moderately successful throughout the 1980s but during the 1990s it boomed. The catalyst to the massive upsurge in business was the clampdown by the authorities on illegal timber shipments to Surabaya and Gresik (see Chapter 4). This clampdown on undocumented timber at Surabaya and Gresik left a clear market advantage for the Telaga Biru entrepreneurs, none of whom were bothering to obtain transport permits for their cargoes.

In this advantageous economic situation, Pak Yakob himself became the principal organiser of the local timber trade, liaising with all the relevant officials to ensure that everything ran smoothly. He was well aware of the legal requirements for transport permits and that these were not being fulfilled: "It was 100 per cent illegal... Of course they [officials] all knew it was illegal but they just looked the other way." Any bribery occurring at this stage was on a small scale. "Everything worked well during Suharto's time", he said. "There was no problem until Habibie became President."

In 2002, with water police and navy patrols demanding 100 per cent compliance with the timber transport regulations, those heady times of the early 1990s seemed remote. Pak Yakob was bitter about the severity of the crackdown on the timber trade on Madura, with his frustration being evident not only towards the central government but also the regional government. In particular, he felt that the kabupaten administration, based in Bangkalan, had done nothing to help the people of the north coast.

Despite his real concern for the ordinary people of Telaga Biru, now severely deprived of employment opportunities, it seemed that for Pak Yakob himself the economic malaise had another negative aspect. Not only was Telaga Biru a much less prosperous place than it had been just a short time earlier but it was also less vibrant. Although still vigorous and physically fit, Pak Yakob was poorly suited to be effective in the now greatly changed political and economic conditions. He had been in his element in the recent past, when the traditional characteristics of individualism and self-reliance were still paramount. Significantly, he made no reference to the possibility that Telaga Biru's current economic woes might have been alleviated if its entrepreneurs had all stood together, as at Sepulu. But for that to have happened, a different style of leadership would have been needed, one based on consensus as well as charisma.

Haji Rofi'i

One individual who was able to offer such a leadership was Haji Rofi'i, the head of the desa of Prancak, at Sepulu (see Chapter 4). In contrast to the forthright and garrulous style of Pak Yakob, Haji Rofi'i's manner during interviews in 2002 was urbane. He was about 40 years of age, a large man but not overweight, articulate with excellent Indonesian language skills, and a good listener as well as speaker. He appeared a man for the times, able to devise and present constructive arguments and plans for the future and who seemed as though he could be as comfortable and effective in parliament in Jakarta as he was in his desa office. Unlike the traditional notion in Java and Madura that 'knowledge is power' and should therefore be dispensed as sparingly as possible, in order to retain the distinction between the one who knows and the many who do not, Haji Rofi'i preferred that the public be informed. This was evident from the desa office at Prancak, which had an unusually open and friendly atmosphere for such a place, with copies of the daily newspaper from Surabaya available for anyone to read.

Visiting Haji Rofi'i at his house, it was immediately apparent that he was wealthy. His house had reportedly cost over one billion rupiah, a huge amount in that local economy. It was large and elegant with wide ceramic-tiled verandahs all round but only a single storey, eschewing the usual aspirations of the new rich in Indonesia to vaunt their success through the conspicuous display of a showcase multi-storeyed dwelling.

For most locals, Haji Rofi'i seemed to be one of those rare individuals for whom everything goes right. He had a good family background, being the son of the former longtime desa head, Nur Habib, who had died in 1998. Nur Habib's income as desa head was only modest, but he was also an extremely successful maritime entrepreneur and fish trader. According to a popular and apocryphal local story, Haji Rofi'i's phenomenal good fortune derived from a talisman, a red stone that had belonged to his father. According to the story, many years ago his father had a dream about a bird excreting a red stone. He related this to one of his crew, who replied that he too had had the same dream. Some time later, this crew member happened to see, to his astonishment, a bird pass out a red stone. He at once picked the object up. Remembering the dream and that his boss, to whom he was much indebted with gratitude, had had the same dream, he brought it to Nur Habib and gave it to him. From that day on, so it was said, the father, and in turn his son, were blessed with good luck.

As a young man, Haji Rofi'i had been fortunate enough to receive a thorough education, studying at an Islamic college in Java. Then he was fortunate in marriage, his bride being none other than the daughter of the richest man in Sepulu, who had made his wealth from fish trading. Haji Rofi'i went into fish trading himself, as his father had done. He was not afraid of hard work and went to sea in his own *golekan* on extended voyages, buying fish from vessels at sea off Masalembu and the Sembilan islands, near the coast of South Kalimantan. The fish would be salted down at sea and on return to Sepulu would be set out to dry for one or two days before being trucked to Surabaya.

When fish wholesaling started to decline in the early 1980s after entrepreneurs from Kalimantan had become involved, Haji Rofi'i turned to copra, buying this commodity in Central Kalimantan and selling it in Madura. Unfortunately the local copra market was then being flooded by copra from nearby Bali, so after only one year he gave up on copra to try his hand as a timber trader. The timber business in Sepulu was then on a very small scale. Because he did not own a sufficiently large vessel himself, Haji Rofi'i worked as a skipper on one of his father's vessels, purchasing his own cargoes of timber and selling them for a profit in Sepulu. Meanwhile, the freight earnings from the carrying vessel went to his father.

At the time of his death in 1998, his father, Nur Habib, owned five prahus. All were inherited by Haji Rofi'i, who was then the leading timber trader at Sepulu. The following year Haji Rofi'i was elected as the head of Prancak desa. He brought sound leadership and, together with his colleague Haji Komar in the neighbouring desa of Sepulu, was the main force behind the Sepulu Entrepreneurs' Association, which ensured the survival of the local timber trade after the government had clamped down on illegal timber imports and revenue evasion on Madura (see Chapter 4).

As well as proving himself an able administrator and businessman, Haji Rofi'i was also a competent seaman. Moreover, he had acquired a sound understanding of the principles of hull design and machinery for engine-powered vessels, setting him apart from most other Madurese prahu owners. He spoke easily, with knowledge derived from experience rather than from books, on matters of propeller diameter and reduction gearboxes and their effect on performance. This thoughtful approach was evident in his new vessel, *Gunung Mas* (Golden Mountain), which was launched in mid-2002. In contrast to the enlarged vessels of the *golekan* type, which were too blunt in the bow and too fine in the stern to make efficient motor vessels, *Gunung Mas* had a modern hull form with sharp entry, full-bodied 'ship-form' mid-section and a broad stern. The engine was an eight-cylinder, ex-truck Nissan RD8 diesel with a 1:4 reduction gearbox.

For all his good fortune and intelligent approach, however, Haji Rofi'i knew well that fate can deal a rough hand to maritime entrepreneurs. By early 2002 two of the five vessels that he had inherited from his father had been lost. One had caught fire and was gutted in port at Sepulu under unknown circumstances while the other, an old vessel that was a chronic leaker, sank in heavy seas in February 2002 near the western entrance to Madura Strait. Fortunately, all crew members were rescued. These losses were significant setbacks but an entrepreneur of Haji Rofi'i's resources could weather them. Then, in December 2002, on only her third voyage, *Gunung Mas*, the pride of the Sepulu fleet, burned and sank. Worse still, this catastrophe occurred on the homeward leg, resulting in the loss of over 200 cubic metres of timber. The fire occurred within sight of the coast of Madura and the crew members, who had dived into the sea, were saved by a local fishing craft.

The cause of the fire was not clear. What was known was that when the vessel was just a couple of hours away from Sepulu, the engine had stopped because the fuel tank had run dry. These prahus are not usually equipped with fuel gauges and the crews rely on experience and manual checking of fuel levels. Nevertheless the situation was not serious because extra fuel was carried in a separate drum. In due course fuel was flowing again and the engine was restarted. However, about an hour later one of the crew noticed smoke coming out of the timber-filled hold. When the men lifted the large and heavy hold hatch, they were met by a conflagration that was beyond their capabilities to control, there being no fire extinguisher on board. The end was inevitable. The skipper was so stricken with shame and remorse that he left Sepulu shortly afterward to seek work in Malaysia. Regardless of whether or not Haji Rofi'i blamed him, it would have been very difficult socially for the unfortunate skipper to remain in Sepulu under the circumstances. For Haji Rofi'i, the sinking of the *Gunung Mas* was a massive blow because neither vessel nor cargo was insured. His personal loss would certainly have been in excess of Rp 200 million.

Haji Matsahri

Haji Matsahri is a highly respected prahu operator from the desa of Sreseh, on the south coast of Sampang (see Map 8). A tall lean man, he was still physically tough enough in 2002, in his late fifties, to pull himself up the flaring sides of an empty prahu, riding high in the water, by a single rope. He then owned one medium-sized vessel which was bringing him a good living. He was one of the wealthiest people in the local community, although his wealth would have been modest compared to that of Haji Rofi'i. Devoutly religious, he had always tried to live according to the principles of Islam, while being tolerant of others who did not maintain such standards.

He was born in the hamlet of Batuputih, Sreseh, the second son of a man who, after making the pilgrimage to the Holy Land, became known as Haji Hotip. Haji Hotip's own career was remarkable. Originally a fisherman from the village of Labuhan in Sreseh, he had married a woman from Batuputih, near the entrance to the Blega river. As is usual in Madura, the young husband relocated to live with his wife's parents, working as a fisherman from his own *sampan*. After a few years of hard work and frugal living, he managed to buy enough

Photo 39 Haji Hotip's first vessel, *Si Baru Dateng*, in the Sampang river, 1983.

timber to build a small *janggolan*, hiring a local man named Haji Umar to help in the work. The vessel was of modest size, in keeping with her owner's means, but she was extraordinarily graceful for Haji Umar was a superb craftsman. Haji Hotip gave her the appropriate name of *Si Baru Dateng* (The Newcomer).

With *Si Baru Dateng*, Haji Hotip prospered from the shipping of salt and built a second small vessel, *Bintang Samudra* (Star of the Sea), again employing Haji Umar to help with the construction. He then built another vessel, *Baru Hidup* (Newly Living). Unfortunately, on its maiden voyage this vessel, heavily laden with salt, encountered rough weather coming out of Sampang and sank after a wave washed over the side and into the hold.[1] The vessel was recovered at low tide without serious damage but the cargo of salt had of course dissolved. Haji Hotip was not liable for the loss of the cargo, which had to be borne by the salt owner, the then head of the desa of Pangarengan

[1] Small *janggolan* for use only in the Madura Strait were usually undecked forward to facilitate the loading and unloading of salt (see Photo 17).

Photo 40 *Bunga Bhakti* sailing hard towards Beliton, 1986. The vessel is floating high because it has no cargo or ballast. The flat object extending beyond the stern is the starboard rudder, stowed on deck.

on the eastern side of the Blega estuary (see Map 8). Nevertheless, Haji Hotip sold *Baru Hidup* immediately after this incident, fearing that the vessel might be unlucky.

From the proceeds of this sale, Haji Hotip built a much larger vessel, *Kota Mesir* (City of Egypt), which was subsequently used to carry salt to West Java. He later exchanged this for a smaller vessel, *Sinar Jaya* (Glorious Light), with a cash payment to make up the difference. With this cash payment plus his shipping earnings, Haji Hotip then built two more vessels, *Fajar Islam* (Dawn of Islam) and *Bunga Bhakti* (Loyal Flower), of 65- and 120-tons capacity respectively. *Bunga Bhakti* (Photo 40) eventually became the best-known vessel of the entire *janggolan* fleet and brought considerable prosperity to Haji Hotip and his family from both the salt and timber trades.

Throughout the 1970s and 1980s, Haji Hotip spent most of each sailing season in Kali Baru in Jakarta, where eventually he set up with a second wife, a Sundanese woman, in a relationship that was never fully accepted by the rest of his family. For business purposes he often commuted back and forth between Jakarta and Madura, always travelling by bus for economy. He built three more vessels, all of moderate size, for the salt trade to Pasuruan, and named them after bus companies that he used for his journeys to and from West Java: *Maju Makmur* (Prosperous Progress), *Sari Indah* (Beautiful Essence) and *Sinar Hati* (Light of the Heart). All these vessels were *janggolan* for the *sepel* model had not yet been introduced at Sreseh.

Haji Hotip's sons thus grew up in a milieu steeped in boatbuilding, seafaring and trading. From an early age, the second son, Daruji, who would later become Haji Matsahri, was lent a hand in prahu construction, learning from his father and the peerless Haji Umar. He left school at 12 years of age and thereafter worked around the village helping his father, until as a young man he started work fulltime as a boatbuilder. He later went to sea on some of his father's vessels but never as a skipper. At this stage of his life he was overshadowed by his elder and more ambitious brother, Haji Zaini, who progressed from skippering one of his father's prahus to become a timber trader. Instead, Haji Matsahri set up on his own behalf, buying an old prahu, *Bintang Timur* (Star of the East), from Pangarengan on the opposite shore of the Blega estuary. He repaired and enlarged this vessel, then operated it himself for one year as a salt carrier before selling it to a man from Pamekasan. After taking the vessel on a 50 per cent deposit, the new owner unfortunately became involved in an expensive divorce. Unable to meet his financial commitment, he abandoned the prahu in a creek at Gresik Putih, near Kalianget in Sumenep. Haji Matsahri later hired a motor vessel, at considerable expense, to tow his prahu all the way back to Sreseh. He knew it was a waste of money to do this because the vessel was then filled with mud and in a sorry state. Nevertheless it was his first prahu and his heart was heavy for the vessel itself, a thing of soul which he had created with his own hands. "It was only like that that I could have peace of mind", he related. *Bintang Timur* was laid up in Sreseh and never sailed again.

As he matured, Haji Matsahri became known as a man of integrity. Coupled with his spirit of enterprise, his good reputation enabled him to establish a niche as a salt agent, liaising between salt producers and prahu owners and buyers in Pasuruan and Surabaya. He so prospered that by the early 1990s he was able to build a large vessel for himself, *Ibu Dana Besar* (Mother of Great Wealth, but usually referred to as just *Dana*). He used this vessel in the Kalimantan timber trade, carrying timber to Pasuruan. So successful was this venture that one year later he established his own timber warehouse, with his eldest son Marthawi as resident manager. In this respect, Haji Matsahri was following his elder brother Haji Zaini, who already had his own timber warehouse in Pasuruan. In 1997 Haji Matsahri sold *Dana* to a man from Sepulu, then built his latest vessel, *Sumber Makmur* (Source of Prosperity).[2]

Although he claimed that boatbuilding was for him a secondary occupation, being at heart a trader, Haji Matsahri was well known for his conscientious approach to building. Consequently, his vessels consistently sold for higher prices than those built by others in his locality. Yet although he enjoyed the challenge of vessel building, for him such work was a means to an end rather than an art form and he readily admitted that his craftsmanship was not in the same class as that of his late mentor, Haji Umar. Indeed, with the increasing emphasis on economic rationalism in vessel construction since the late 1970s, it is unlikely that anyone in Madura will ever again reach Haji Umar's standard.

Haji Hotip died in 1999, aged 74. He had risen from humble beginnings to become the most successful prahu operator in the Sampang area while his two eldest sons had also become successful entrepreneurs in their own right. By the time of his death, however, Haji Hotip's fleet had dwindled to just one vessel. *Bunga Bhakti*, regarded as almost a member of the family, had been lost after running aground in the Karimun Jawa islands (see Map 5) a few years earlier, while several other vessels were thereafter either similarly lost or sold. The only surviving vessel was the smaller *Sinar Hati*, originally built for the salt trade to Pasuruan but by this time working as a timber carrier supplying Haji Zaini's warehouse in Pasuruan. After Haji Hotip's death, *Sinar Hati* became the property of his widow, with the freight profits flowing to her. Although in her mid-seventies, this lady was in 2002 regularly undertaking the long journey by car, ferry and bus from Sreseh to Pasuruan, staying at Marthawi's warehouse for a few days before returning to Sreseh. In this way she could enjoy the company of the members of her family in Pasuruan while keeping herself informed on the running of the business.

---

[2] *Dana* was substantially enlarged in Sepulu by her new owner to carry over 200 cubic metres of timber. Unfortunately in January 2002 this vessel, fully laden with timber, sank in rough conditions off Probolinggo in the Madura Strait and was a total loss. All the crew were saved.

Haji Matsahri had two younger brothers who were also living in Batuputih. Together with their mother, the family matriarch, the three brothers and their immediate families all stayed in the family compound, in a series of closely spaced dwellings after the traditional pattern. By 2000, however, the main dwelling in the compound, occupied by Haji Matsahri and his wife and children, had fallen into a rather dilapidated state. The bamboo mat ceiling was damaged in places, with roof tiles visible through the rents, while the thin plastering of the floor was badly cracked and chipped away. Just ten years earlier such less-than-pristine conditions would have been of no consequence, provided that the dwelling was serviceable, but since then many people in Sreseh had built new houses clad with ceramic tiles using earnings from Saudi Arabia, causing a sense of embarrassment among some families owning older plain houses in run-down condition. Said Haji Matsahri in 2001: "Just look at this place [the interior of his house]… I feel ashamed whenever guests come." Although he had never coveted high social status, he was by then aware that he was widely regarded as a successful entrepreneur and a pillar of society, with all six of his children having been educated in *pesantren* and two now working in Saudi Arabia. With the sumptuary standards in the community having risen, it was appropriate for him to build a new house commensurate with his social standing.

This social pressure was too much to resist and by early 2003 Haji Matsahri had a new house under construction, just across the road from his existing house. The new house was an imposing edifice, with its brick walls four metres high to give relief from the heat, while the yard enclosure was set off in the front by an impressive 1.5-metre high barrier fashioned from 75-mm diameter stainless steel pipes, a type of structure popular with the wealthy families of Sepulu. The house was divided into two equally sized separate living quarters, one for Haji Matsahri and the other for his eldest son Marthawi, who for the past few years had been living with his wife and children in cramped quarters at the back of the Pasuruan warehouse.

Although in early 2003 the house was still far from finished, it was clear that Haji Matsahri was already enjoying the prospect of the comfort and prestige that it would provide. Yet he did not take his prosperity for granted, being keenly aware of the vicissitudes of maritime entrepreneurship. Indeed, in April 2002 a vessel that he had built for his younger brother, Haji Cholil, had been lost due to fire only a few months after its launching. This was a financial disaster not only for Haji Cholil but also for the eldest brother Haji Zaini, who as 'timber boss' lost a consignment worth Rp 25 million. The vessel had just finished loading at Marabahan on the Barito river (Map 13). Although the cause of the blaze was not clear, it started in the early evening after a small petrol-powered generator, used for lighting while in port, had been started up on the deck. According to reports, the deck burst at once into flames. There was no fire extinguisher on board.

Despite this calamity, Haji Matsahri had not installed a fire extinguisher on his own prahu. Like most Madurese prahu owners, he knew little about such devices and did not know where they could be bought. However, he did intend to raise the freeboard to increase the carrying capacity. And every year at the *Ketupat* festival, held one week after the end of the fasting month, an occasion when livestock and vessels are blessed and feted, he would arrange for the local kiai to come on board and bless *Sumber Makmur* and say prayers for her continuing safety and bountifulness.

Pak Sum

Pak Sum was a prahu owner from the desa of Gedugan, in the eastern part of the island of Giligenting (Map 9). A small wiry man, he was about 70 years old when interviewed in 2003. Although everyone around him referred to him only as Pak Sum, this was not his own given name but an abbreviated reference to his firstborn child, his daughter Sumaniyah. 'Pak Sum' thus meant not 'Mr Sum' but rather 'Sum's father'. This self-effacing term of address is traditional among common folk in much of Indonesia, although in Madura the custom has now faded out so that it is unusual for men of less than middle age to be so addressed. Pak Sum was thus a 'little person'. None of his five children were educated in *pesantren* — "it would have been fine, but where would the money have come from?" — and of course to make the pilgrimage to the Holy Land was for him just a dream. Yet while he accepted his place in society as an ordinary householder, he was in his own unassuming way remarkable, the sort of person for whom the phrase 'salt of the earth' is particularly apt.

Although there is a large strand village at Gedugan, with many transport vessels and fishing boats operated from there, Pak Sum's house is about two kilometres inland in a tiny hamlet called Ragang. The sea is not visible from Ragang and the hamlet appears to be purely agricultural, surrounded as it is by small maize fields and coconut palms. But the agricultural economy on Giligenting has always been marginal and for most of his adult life Pak Sum has looked to the sea to supplement his earnings from the land. The dispersal of his own family was a testament to the seafaring way of life on Giligenting. Of his five children, only Sumaniyah remained on the island, in the same family compound at Ragang. The others, two sons and two daughters, had all moved away — not to the city of Sumenep or even the mainland of East Java but to distant Cirebon in West Java. Cirebon has long been an important port for Giligenting vessels and a substantial expatriate community from the island has become established there. Both of Pak Sum's sons were formerly prahu sailors, as were the husbands of his two daughters. None of his extended family in Cirebon worked as sailors any longer, however, with all of them having switched to the food stall (*warung*) trade for their livelihoods.

Pak Sum started going to sea during the 1950s on a *lete-lete* carrying salt from Madura to ports in West Kalimantan: Ketapang, Pontianak, Singkawang and Pemangkat (see Map 7). The salt was loaded not only from nearby Kalianget but also from Bunder in Pamekasan and Sungai Rajo (the Blega estuary). In those days the vessels were mainly of modest size, typically carrying about 35 tons of salt. As well as salt, the vessels Pak Sum worked on often carried horses from Bima on the island of Sumbawa, or Ampenan on Lombok, to Pasuruan or Tegal.[3] The boats could carry 25 to 30 horses, and would reach Pasuruan in 6 days or Tegal in 8. By comparison with Madurese cattle and their skittish nature, the horses were easy to manage. "Those horses just walked straight up the [gang]plank and on board, as if they were used to it", he said. This trade in horses ceased after the 1950s, while the salt trade lasted until the 1970s. By then timber had become the major cargo.

Unlike the vessels from Telaga Biru, which would return home several times during the sailing season, the Giligenting vessels usually stayed away for the duration of the southeast monsoon. Communications with home were tenuous. In those days there were no telephones on Giligenting and letters were rarely used as many of the sailors were barely literate. Such news as was conveyed was usually by word of

[3] These horses were smaller than the standard European breed of horse. They were used for pulling small carts (*dokar*) for the carriage of passengers and goods. This method of transport was still very common in Madura and in outlying parts of Java.

Photo 41 Furling the mainsail on a Giligenting *lete-lete*, with six crew members perched on the yard, 1984. The prahu will enter port under the small foresail alone, which can be let right forward of its mast to spill the wind as required.

mouth from any sailor who happened to come back during the sailing season. Only rarely were such tidings sorrowful, however, because the *lete-lete* were seaworthy and the crews hardy and agile. To a Westerner, the task of furling the mainsail, with up to six men clinging on to the steeply sloping yard (Photo 41), might seem dangerous but Pak Sum had never heard of an accident doing this work.[4]

In those days none of the vessels carried compasses or charts but they usually reached their destinations without difficulty, relying on crew experience and the landmarks of the Java Sea. The main danger was sailing too close to land or small islands at night. Indeed, Pak Sum had himself twice experienced shipwreck during the hours of darkness. The first occasion was in 1983, when a *lete-lete* he was crewing on ran aground on a small and uninhabited sandy islet to

[4] The Englishman G.W. Earl, who had considerable experience with Javanese crews on European sailing ships, had a high opinion of Javanese sailors. Concerning their agility when working aloft, Earl stated that in this respect the native crewmen were "inferior only to monkeys" (Earl, 1971 [1937]: 84).

the north of Cirebon with a full load of timber. It was five days before they were rescued but the vessel was intact, albeit unusable, and they had water and rice on board. The second shipwreck occurred in 1986, when he was sailing on a *lete-lete* bound for Airhitam, on the southern coast of West Kalimantan (see Map 15). Sailing fast before the southeast wind, in the darkness the empty vessel smashed ashore on one of the Karimun Jawa islands and was totally wrecked. In another incident in the family, Pak Sum's elder brother was also wrecked at night time: en route from Pasuruan to Madura with a load of bagged cement and concrete blocks, the vessel struck a reef off the island of Mandangil, south of Sampang. In all three of these shipwrecks, the crews were saved. However, in 1993 a *lete-lete* skippered by a man named Sahut left Giligenting and was never heard of again — "maybe run over by a ship, maybe hit a rock... Who knows?" said Pak Sum. In another tragic case in 2001, a motor vessel from Giligenting departed Kumai bound for Java with 21 passengers and a cargo of timber but no trace was ever found of the vessel or the people on board. Such stories were, fortunately, rare. Pak Sum was of the view that the newer vessels, with an engine and at least a modicum of navigational equipment — a compass and a chart — were, all things considered, safer than the vessels of the past.

Despite his many years at sea, it was only in 1997 that Pak Sum became a prahu owner. Lacking both the capital and practical skills to build a new vessel himself, he bought a second-hand one, *Sumber Rezeki* (Source of Good Fortune), that had been built in Giligenting in 1978. It was originally rigged as a *lete-lete* but was subsequently modernised with a larger aft deckhouse and a sloop rig. Being engineless and rather old when Pak Sum bought it, this prahu was not very expensive, especially because with a capacity of only 120 cubic metres of timber it was considered too small to be profitable without a trading interest. Pak Sum lacked the capital to become involved as a trader but he did know a man on the north coast of Madura who wanted to invest in importing timber and who trusted him. Although his return from freight on this vessel was modest, it was adequate for Pak Sum's needs. Nevertheless, frugality remained a way of life for him. With this philosophy he had never been tempted to install an engine. Apart from the fact that he knew little about engines, *Sumber Rezeki* was too old to warrant such an additional investment, which would reduce rather than increase his profits and possibly shorten the life of the vessel too if the plank seams opened up from the vibration.

With his venture into vessel ownership proving successful, in 1999 Pak Sum bought a second prahu, *Dharma Bhakti* (Loyal Duty). Slightly smaller than his first vessel, *Dharma Bhakti* had been built on Sapudi but was subsequently bought by a man from Galis, on Giligenting. Before buying this vessel, Pak Sum discussed the matter with his timber boss, who in the event had been keen to expand his own business. Two years later, however, *Dharma Bhakti* was arrested with a load of undocumented timber and towed to Kalianget. The vessel was later released but the skipper, Pak Tolak, a nephew of Pak Sum, spent three months in prison as a result of this affair. After this experience, Pak Tolak, who was in poor health, ceased going to sea. His family was not well off financially and dependent on the earnings of his 16-year-old son, who was then working as an ordinary crew member on *Dharma Bhakti*.

During the northwest monsoon, both of Pak Sum's vessels would return to Giligenting. Although the most popular place around the island for sitting out the wet season was along the main beach at Gedugan, Pak Sum preferred to lay up his prahus at an isolated spot on the other side of the Gedugan peninsula. Facing towards the southwest, this place was sheltered from the worst weather of the northwest monsoon. Nevertheless, it was not used by other vessels because the coast there was littered with coral outcrops, while the beach was harsh and strewn with broken coral, unlike the sandy beaches around the greater part of Giligenting's coastline. For Pak Sum, however, it had the advantage that it was only about ten minutes walk from his house, enabling him to carry out much of the maintenance work himself at his own convenience.

While life seemed uncomplicated in Pak Sum's hamlet, with no electricity or piped water supply, it was far from idyllic. His house was adequate but basic, like his vessels. The family still relied on kerosene pressure lamps for lighting and retired to bed early. There was a well, always a blessing to have in rural Madura, although water had to be drawn up by hand for want of a pump.

Although he no longer went to sea, Pak Sum had not retired from active work, nor did he have any intention of doing so while he was physically capable. Much of his time was occupied with tending two cows, which were kept in a covered pen alongside his house. These cows belonged to someone else but Pak Sum tended them so that later, when the animals were sold, the profits would be shared between him and the owner. Cattle-rearing, the traditional standby for rural Madura, had always been important in Pak Sum's household economy. He had his two prahus, largely paid for in the first place through cattle-rearing, but one never knew what might happen at sea. Cutting fodder for cows is a big chore, especially during the dry season when the animals cannot forage for themselves near the house. Pak Sum did this work himself, squatting and crawling on the ground while using a sickle, the ubiquitous agricultural tool of Madura. When he had gathered a large enough pile, he would tie it in a huge bundle to carry back home on his head. Few outsiders, if they saw him working away briskly like this, would have guessed that he was the owner of two prahus engaged in the Kalimantan-Java timber trade. But with his diligence, thrift and self-reliance, Pak Sum represented the very qualities that have made the Madurese such a force in indigenous shipping.

Haji Amir

Haji Amir Hosein, a vessel owner from the small town of Gayam on the island of Sapudi, was in 2003 the most successful maritime entrepreneur in all Madura. He owned five vessels, all large. As it happened, probably the second most successful vessel owner in all Madura was also from Gayam, being his friend Haji Ariyono, a dapper local official who owned three large vessels plus another under construction. While there were many more vessels on Giligenting than on Sapudi and a few Giligenting owners with three or four vessels to their names, in terms of tonnage these two entrepreneurs from Sapudi were in a class of their own.

In contrast to the style of most wealthy people in Java and now also in some parts of Madura, especially along the north coast, there was nothing about Haji Amir's person or manner to suggest that he was an outstandingly successful businessman. His house, like that of Haji Ariyono, was comfortable and well-appointed but not at all ostentatious, nor did he own a car. A robustly built man in his early sixties, he spoke in a steady, direct manner. Perhaps the most remarkable feature about him, as became apparent after listening to him for some time, was the objectivity and dedication of his approach to his shipping business.

It was little surprise to learn that Haji Amir's late father was a prahu operator, owning a large 50-gross ton *lete-lete* called *Jindrawasih* (Bird of Paradise). However, his grandfather was a farmer, while Haji

Photo 42 *Merpati* (*Madura Bahari Bersama*) at Semarang, 2003.

Amir's only brother was a teacher, now retired. When his father died, Haji Amir, as the seafaring son, inherited *Jindrawasih* and he acknowledged that this start was vitally important to his later success.

With the profits from operating his late father's vessel, in 1977 Haji Amir built a vessel for himself on the beach at Gayam, giving it the rather prosaic name of *Nusantara Bahari I* (Maritime Archipelago I). This was also a *lete-lete* hull but it was equipped with an engine, unlike *Jindrawasih* which was a pure sailing prahu. This new vessel was followed over the next few years by three more: *Nusantara Bahari II*, *Nusantara Bahari Bersama* (Our Common Maritime Archipelago) and *Madura Bahari Bersama* (Our Common Maritime Madura). As these names were rather awkward for conversational purposes, each vessel also had its own 'pet' name. *Madura Bahari Bersama*, for example, had always been referred to colloquially as *Merpati* (Dove). These vessels all had the traditional *lete-lete* hull form but were powered by engine alone. The decision to dispense with the sailing rig was mainly because of their sheer size: *Merpati*, for example, carried 400 cubic metres of timber. At this size the *lete-lete* rig, in its standard proportions, would have required an enormous yard and been almost unmanageable.

While these vessels all performed well, in the next few years after the launching of *Merpati* Haji Amir began to consider the merits of the so-called *pinis* hull form used by the Bugis. In particular, he thought that the *pinis* hull shape, with its relatively drawn-out bow and stern, had the advantage of needing less strongly curved and twisted planks towards the ends than the fuller-ended *lete-lete*. For a given capacity, it should therefore be more economical to build than the *lete-lete*. In this respect Haji Amir was on his own, because at that time no one else on Madura had emulated the *pinis* form. Although a few large non-traditional vessels had been built at Telaga Biru, these were based on the typical small wooden motorships of the *Lokal* fleet rather than the vessels used by the Bugis. In his own words, "I was learning and thinking for myself."

Haji Amir's thoughtfulness became evident in his approach to his next vessel. Generally speaking, if a prahu owner wished to build a vessel of a new model, he simply instructed his workmen to achieve the desired shape to the best of their ability. Haji Amir's existing vessels had all been built like this, relying on tradition. Now, however, he took his own path. When he had acquired sufficient capital, he purchased a load of ulin timber from Batulicin in Kalimantan and shipped it to Sapudi. This was in itself unusual because vessels on Sapudi were traditionally built of teak, which was available from nearby Kangean. Haji Amir then traveled to Banjarmasin, where he engaged an experienced Bugis shipbuilder named Pak Mahasar. He even chartered a small aircraft to take him and Pak Mahasar to Kota Baru in South Kalimantan, from where they travelled by speedboat to Batulicin. For a whole month he observed the shipbuilding at Batulicin and discussed at length with Pak Mahasar the advantages and disadvantages of different styles of vessel. When he was satisfied with what he wanted, he travelled back to Sapudi, bringing Pak Mahasar with him, and set about building his new vessel. It resembled a *pinis* except that the mid-section was more like that of a *lete-lete*, with little deadrise (or 'v-shape') in the lower planks close to the keel, and thus a better form for cargo work. Nor was it built in the same way as a *pinis*. The Bugis teams in Kalimantan would bend the planks around fixed frames with brute force, using clamps and chains,[5]

[5] This was not the traditional approach to building in South Sulawesi, where builders hewed short baulks of timber to achieve the desired plank shapes and assembled the hull shell-first (see Horridge, 1979b). This method of construction is still used in South Sulawesi. Vessels so built are strong enough for normal use but in large sizes they are very vulnerable to hard grounding.

whereas Haji Amir's team, under the supervision of Pak Mahasar to ensure the right shape, built the hull shell-first with the planks pre-bent to shape over fire in the Madurese tradition. Similarly, instead of the flush deck preferred by the Bugis and now also by builders of new vessels at Giligenting, Haji Amir retained the raised deck structure common to Madurese prahus to maximise internal capacity.

This new vessel proved very successful and eventually was followed by three more built along the same lines. With his shift away from the traditional style of hull, Haji Amir apparently felt less constrained in his choice of names for his new vessels, naming them, in order of construction, *Bandung Madura* (Partner of Madura), *Garuda Madura* (Great Bird of Madura), *Tanjung Barcelona* (Barcelona Point) and *Teluk Bayur* (Bayur Bay, *bayur* being a species of tree). Both this set of names and the set used by Haji Amir for his earlier vessels stood outside the usual practice in Madura, where owners typically chose flower names, Islamic names or names suggesting handsome profits or good fortune.

Haji Amir's latest vessel, *Teluk Bayur*, with an expected capacity of 800 cubic metres of timber, was still being fitted out in 2003. In the construction of this vessel, Haji Amir again showed uncommon thoroughness of approach. Because of the large size, and especially taking into account the installation of the engine, he decided to build it at Gresik Putih, Kalianget, rather than on Sapudi. He recruited a team of workmen from Sapudi and brought them to Gresik Putih, where they lived under a tarpaulin shelter. By early 2003 they had been living like this for 18 months. Haji Amir did not stay with them, having plenty of other matters to attend to. When the hull and deck were finished, the vessel was towed to Kalianget harbour for fitting out. Haji Amir then moved into a cheap hotel in Sumenep so that he could oversee the work, spending all day long, every day, at the vessel. He expected that he would be involved like this for at least three months. The capital outlay for this large vessel was expected to come to between Rp 600 and 700 million, which was significantly less than the cost of building it in Kalimantan. According to Haji Amir, to build a hull with a capacity of 500 cubic metres, a Bugis team would use 100 cubic metres of timber, whereas he would require only 75 cubic metres. This economy was largely because pre-bending the planks over fire allowed less wastage of timber than would otherwise be the case.

Photo 43 A large *lete-lete*, possibly *Merpati*, under construction on the beach at Gayam, Sapudi, 1983.

Photo 44 The same beach in 2003. The hull form of this vessel is based on the modern Bugis type. The hull planking is about two-thirds complete.

Like most other Madurese owners, Haji Amir had bought all the engines for his vessels as second-hand truck or earth-moving machine engines from Singapore, through an agent in Surabaya. This was much cheaper than buying a new marine engine such as a Yanmar or Kubota. After all the reconditioning and conversion had been done, the total cost of the engine usually worked out at between half and one-third the cost of a new factory-built marine engine. Maintenance was also much cheaper because spare parts for standard marine diesels were more expensive than parts for converted truck engines. "I buy the engine without seeing it", said Haji Amir. "After getting it checked over, I decide what needs to be done." The engineering work was carried out in Surabaya.

Despite the large capital outlays, none of Haji Amir's vessels have ever been insured. He did lose one vessel, which sank with its cargo in waters to the south of Pontianak in 1978, but he learned from that experience. Since then three of his vessels had been holed after striking rocks, but on each occasion the ship had been saved by patching the hole with leather secured with nails and using a compressor to operate a powerful mechanical pump. By comparison, most Madurese prahus carry no emergency kit apart from some shredded bamboo (*receh*) for caulking.

Apart from a dedicated emergency kit, proper maintenance is also important. In this respect Haji Amir was again ahead of most other Madurese vessel owners. For most prahus, maintenance work is usually done on a beach. However, Haji Amir's vessels were all hauled out once a year in a dry dock, preferably in Banjarmasin, which was the cheapest place, but sometimes in Semarang. The cost of using a dry dock was in 2003 around Rp 3 million, an amount that would deter many small prahu owners. Nevertheless, for such large vessels this charge represented good value because maintenance in a dry dock was far more efficient of time and labour than on a beach. Instead of the usual coating of *kapur* (slaked lime mixed with vegetable oil), Haji Amir used anti-fouling paint to protect the lower part of the hull. His oldest vessel in 2003 was *Merpati* (Photo 42), still with all her original planking intact after 20 years and probably good for many years to come. *Merpati* was planked with teak, but the ulin-planked vessels should be more durable still, with a working life of up to 40 years, barring accident.

Because of their large size, Haji Amir's vessels were usually untroubled by heavy weather and they worked all year round, instead of being laid up for four or five months during the northwest monsoon like smaller vessels. Each of his vessels was equipped with a satellite navigation device, a rare item on Madurese prahus, as well as the necessary charts. This was an important investment for very many prahus have been wrecked due to haphazard navigation. To further minimise risk, on each of his vessels there was a crew member competent in the maintenance of marine diesels and capable of taking the engine apart at sea if necessary. Also, every one of his vessels carried at least one fire extinguisher.

When I spoke with Haji Amir in 2003, he was not sure if he could raise sufficient funds to cover the cost of the ten-cylinder engine for his new vessel, *Teluk Bayur*. This was not a serious obstacle to his venture, however, because he had plenty of business acquaintances in Jakarta, Semarang and Kumai who were all keen to invest with him if the opportunity arose. In a line of business where disaster could strike unexpectedly and where numerous lesser players had withdrawn from the field in recent years, Haji Amir was a good man to have as a partner.

Abdul Majid

A prahu shipping agent based in the port of Juwana, Central Java, Abdul Majid was in 2003 in his mid-thirties. When I first met him on the dockside at Juwana, I was unaware that he was a shipping agent. Indeed, with his wiry physique, casual garb and unassuming manner there seemed nothing about him to distinguish him from the many prahu sailors and stevedoring workers in the port. As we struck up a conversation, however, it soon became apparent that my initial impression of him had been wrong.

He was born in 1967 in Telaga Biru, on the north coast of Madura. His father owned three *golekan* and was also a livestock trader, shipping cattle to Jambi in Sumatra. The family was moderately well-off but not wealthy. As a Telaga Biru lad, Abdul Majid was already unusual in that he had no intention of following his father's vocation, aspiring instead to become a government official. Bright and a high achiever at school, he went on to complete his secondary schooling, whereas most of his peers did not study beyond elementary level. Despite his intelligence and earnest approach, however, his hopes of getting work in the government bureaucracy were not fulfilled, possibly because his father did not have the right connections. For lack of any other employment, he turned to the standby work of Telaga Biru, crewing on a prahu. Unexpectedly, the experience of going to sea changed his outlook on life. He became aware, for the first time, of the business opportunities available in prahu shipping and also of the shortcomings of his formal education. "When I was at school I just studied, without really thinking. My creativity was stifled", he said. "It was only then [after working at sea] that I realised that my education had done nothing to prepare me for a business career, which needs initiative, boldness, imagination..."

From the start of his maritime career Abdul Majid was fascinated by the challenge of pilotage and navigation. On his own initiative while still in his teens, he went to the Department of Marine Communications in Surabaya and, after making enquiries, managed to purchase a chart of the Java Sea. None of his fellow Telaga Biru sailors had such a chart. By applying himself diligently to both theory and practice, he soon gained a reputation as an uncommonly capable navigator. With this technical prowess backed up by excellent communication skills, he so distinguished himself from his peers that in 1989, at only 22 years of age, he was appointed skipper of one of the largest vessels at Telaga Biru, carrying passengers to Pontianak and Tanjungpinang in Riau, the gateway to Malaysia. Many people from northwest Madura were then travelling to Malaysia to work, in nearly all cases without a permit. They paid Rp 250,000 each, a fare that took them from Telaga Biru all the way to Kuala Lumpur. Abdul Majid himself arranged the necessary graft payments to police and immigration officials in Malaysia — while at the Telaga Biru end he would organise a 'dispensation' from the harbourmaster's office to carry 25 passengers on what was ostensibly a voyage to load timber. A dispensation for this number could stand scrutiny in the port records but in fact his vessel carried 250 to 500 passengers on each outward voyage. A full cargo of timber would be carried on the return leg from Pontianak to Telaga Biru.

One year later, in 1990, he accepted a position as skipper of a Sumenep vessel carrying timber from Sampit to Juwana. In the course of his dealings with the port authorities in Juwana, he so impressed the harbourmaster that the latter invited him to become a pilot for the port. Juwana does not receive large ships, with the only vessels calling there being local fishing craft and cargo prahus. In the past there had been no need for professional pilots as the prahus were usually small. With the increasing size of prahus, however, a need for pilot services had arisen. In 2003 the rule for Juwana was that vessels with a cargo capacity of over 100 cubic metres had to use a pilot. The charge for this service was Rp 2,000 per cubic metre of cargo.

From 1992 until 1996 Abdul Majid worked as pilot in Juwana, along the way marrying a young Javanese woman. Yet, satisfied as he was working as a pilot, he was aware that his financial prospects would be much better as a shipping agent. To become a shipping agent in his own right, however, would have required capital far beyond his resources.[6] Therefore, in 1996 he approached the owner of a shipping agency in Gresik, a Javanese man, and proposed that he open a branch agency at Juwana. The shipping agent agreed to this idea and in due course the branch agency was set up with Abdul Majid as manager.

The Juwana shipping agency was doing well, dealing with ten to fifteen vessel calls each month. Abdul Majid worked alone, with no office staff. He claimed to make only a modest profit from each vessel call, between Rp 500,000 and Rp 1,000,000 depending on the size of the vessel concerned. His role covered much more than just port administrative business as he was also active in the timber trade, liaising between financiers, prahu operators and timber merchants (see Chapter 5), and often arranging outward cargoes of rice and fertiliser and, less often, cattle. As a side venture, and drawing on his experience as a port pilot, Abdul Majid was planning in 2003 to set up business in Juwana as a tug operator. Although most of the prahus coming into the port were adequately powered, manoeuvering the heavily laden craft in the narrow confines of the harbour was difficult. The traditional method of manoeuvering in such confined places was by use of a long bamboo pole but most of the prahus were now too large for this method to work efficiently. For these reasons, tug services had become standard. To take advantage of this opportunity, Abdul Majid needed an appropriate vessel. Knowing what he wanted, he contracted a shipwright to build a tug, 12 metres in length, to his

[6] To open a shipping agency, four requirements must be fulfilled. First, an agency must have at least two vessels in its 'stable'. Although the vessels can be owned by anyone, for administrative purposes the agency is regarded as the 'owner'. Second, a permit, called a SIUP (*Surat Izin Usaha Pelayaran*) must be obtained from the harbourmaster for the port concerned. Third, the agency has to join the port stevedoring organisation or PMKL (Perusahaan Muatan Kapal Laut), with at least one truck of its own. Fourth, and probably the most difficult, the agency must have Rp 250 million in its bank account as working capital.

own specifications. Whereas most vessels engaged in towage in prahu ports are crude utility craft, often old and barely adequate for the task, Abdul Majid's little purpose-built tug was capable of pushing vessels displacing 500 tons or more. It was heavily constructed from ulin timber that he had purchased in Kalimantan and was at the time being fitted with a massive ten-cylinder diesel engine. The total cost was expected to come to about Rp 200 million. While this was very expensive for the size of vessel, it was built to last and should prove a good investment.

Abdul Majid was thus well poised to take advantage of the changing circumstances affecting prahu shipping. He attributed his success to his education — notwithstanding his view that it had not prepared him for a business career — together with his ability to relate to others and develop business networks. These were indeed just the qualities and skills often conspicuously lacking among Madurese people. In this respect, Abdul Majid was unsparing in his criticism of most Madurese maritime entrepreneurs and especially those of his home area of northwest Madura: "Nearly all of them have no schooling beyond Year Three... [they are] ignorant, [they] trade speculatively, waste time and money on worthless pursuits, don't plan ahead..."

He was similarly critical of the lack of technical expertise on Madurese vessels. In most cases their sailing rigs were just token rigs to allow classification as motor sailing vessels (*Kapal Layar Motor*, KLM), which, among other advantages, meant that they did not require a properly trained master or engineer. In practice, however, most of them were entirely dependent on their engines and, given their size, they should really have had trained engine crew. During the period when Madurese vessels relied on sail alone, total failure almost never occurred with the sailing rig because the making and repair of the rig was within the capability of the crews. As a matter of survival, the sailors were extremely resourceful in this respect. With the change to engine power, however, that expertise and resourcefulness had gone, with hardly any formally qualified marine engineers working on Madurese prahus. In the case of Haji Amir's vessels, each of which was supposed to have at least one man knowledgeable about engines, Abdul Majid stated, "They still have no proper training, they just have the guts to take a diesel engine apart and hope that they can put it back together again." At the same time, he was also critical of the abandonment of the sailing rig for smaller vessels. Engineless craft have become marginalised in recent years, mainly because consignors prefer motor vessels, but Abdul Majid regarded this attitude as lacking objectivity. Indeed, he was enthusiastic about the economic advantages for engineless craft as a consequence of the steep rise in the cost of fuel, with some engineless vessels employed through his agency. He even ventured that "the most useful thing about having an engine is that a vessel can disappear quickly" [regardless of wind direction or force] in order to evade a police operation.

Abdul Majid also disagreed with the general pessimism in Indonesia, including in port administration circles, over the future of the Kalimantan timber industry:

> People say that it's [nearly] finished but they just follow what the NGOs say and never look for themselves. Even if they do look, it's just along the main road... I've travelled far into the interior, from Pontianak, from Sampit, from Kumai...driven a car for a whole day along back roads, without running out of forest all around. The timber isn't finished, it's just that so much goes to West Malaysia because the price there is higher than in Java. Even [oil palm and pulpwood tree] plantations can't really threaten the forest, because they can't make plantations on hilly country... I've looked seriously, I'm certain that timber for the domestic market will never run out. The export market is different because only top-quality wood goes for export. Top-quality wood will eventually be in short supply but for the construction industry we can use rougher wood. It doesn't matter if it's not perfect.

Such words, directly against the current of opinion widely expressed in the media and echoed by most port officials, demonstrated Abdul Majid's ability to think for himself and not be swayed by popular notions. This was not blind optimism but a balanced view that took into account the end of old-growth forests and the decline in timber quality.

Unlike most of the well-established shipping agents in Juwana, Abdul Majid did not have an office of his own in the port complex. Instead, he used the front part of his house, situated about a kilometre from the port, as his business premises. The house itself was basic, not at all like the opulent ceramic-tiled residences of the new rich of the north coast of Madura. The interior walls were of simple white-washed plaster, marred by large cracks and scribblings by his two small children. The money spent on the tug vessel could have paid for a grand house but the business investment had to come first, to provide for the future.

With their modest dwelling, he and his wife seemed not much different from many other struggling young couples in Java. Yet in his own way Abdul Majid was extraordinary in that he had succeeded in bridging the gap between the traditional maritime villages of the north coast of Madura and the realities of the shipping business in modern Indonesia.

Between Tradition and Modernity

From the biographical information presented in these six profiles and building on material presented in earlier chapters, it is apparent that the conditions facing prahu operators in the early twenty-first century have changed much since the early period of the timber boom. First, to be competitive *Pelayaran Rakyat* vessels now need to be much larger than those of a few decades earlier. Whereas most *lete-lete* engaged in the timber trade during the 1960s and 1970s were between 50 and 80 cubic metres capacity, by 2002–03 vessels below 150 cubic metres capacity were regarded as too small for freight work. Below this size, to be economically viable, vessels either had to be engineless, in which case it was difficult to get transport work anyway, or the vessel owner had to have a trading interest in the cargo. Motor vessels of 400 cubic metres and above, such as those owned by Haji Amir and Haji Ariyono, were big enough to give good returns on freight and also earned income on both outward and inward voyage legs. Haji Amir's vessels regularly carried cargoes of fertiliser from Gresik to Kumai, while Haji Ariyono's carried the same commodity from Semarang to Kumai. Nearly all small prahus, by contrast, sailed empty on the outward leg.

A second and consequent observation is that by 2003 it was far more difficult than it had been a few years earlier to raise the capital for a new vessel. As exemplified by the case of Haji Matsahri's father, Haji Hotip, before motorisation it was possible to rise from very modest circumstances to become a successful maritime entrepreneur. Since the early 1990s, however, such a leap has become vastly more difficult, so much so that one of the key factors for success in prahu shipping in recent years has been to have a father who was already a successful vessel owner. In the cases of Haji Rofi'i and Haji Amir, the headstart derived from their fathers' businesses is obvious. I also met many other prahu owners who had built upon the success of their fathers.

Apart from the sheer increase in vessel size, the cost of an engine is a burden that vessel owners did not have to contend with in the days of sail. While sailing rigs were not free either, as devised and operated by the Madurese they were very economical and much cheaper than the European-derived sailing rig used by the Bugis. The financial burden of installing an engine became much greater with the collapse of the rupiah in 1997–98, after which all imported components became very much more expensive. The cost of timber suitable for marine work has also increased greatly, well beyond general inflation. As a rough rule of thumb, in 2003 the cost of a new vessel in Madura, including an adequate engine, was between one and two million rupiah per cubic metre of cargo capacity. A 150-cubic-metre capacity vessel, about the minimum feasible size nowadays, would therefore have cost about Rp 200 million, a very large sum in a village economy. This was notwithstanding a strong emphasis on cost cutting, with engines almost always being converted truck engines and fuel tanks made from recycled plastic chemical containers supported by an exterior steel cage, a practice that would certainly not be acceptable in modern industrialised countries.

There were two ways around the problem of raising capital. One was to take on a financial partner as engine owner. This had become fairly common for both small and large vessels but it did, of course, reduce profitability for the vessel owner and was therefore avoided if possible. The other was through credit. However, banks were no more willing to invest in prahus than they were 100 years ago. Instead, credit was still being arranged through informal channels, by borrowing gold from relatives and friends. Sometimes such borrowing was from a large number of individuals, some of whom contributed only small amounts.

Although this village-level mode of organisation clearly places prahu shipping in the informal economic sector, the level of capital investment required is now well beyond that usually associated with the informal sector. At the same time, the industry is still not free from the uncertainty associated with informal-sector enterprises because there is no insurance on the vessels. The absence of insurance reflects the informal nature of this maritime transport sub-sector and the traditional social world of the actors within it. Insurance has never been part of the prahu shipping economy and vessel owners do not seek to have it, despite the large increases in vessel size and investment outlay. For that matter, most prahus, and certainly most Madurese ones, would not come close to meeting the conditions that a reputable marine insurance company would require before issuing a policy. The traditional approach towards 'insurance' among prahu owners has instead been to aim to own more than one vessel, so that if one is lost the owner can still have an income from the remaining vessel(s). Although the increase in size and required outlay has made it more difficult to own more than one vessel, this attitude still prevails.

To conclude, however, that these maritime entrepreneurs do not know how to access insurance, or that it is too expensive for them, would be simplistic. Choosing to 'self-insure' by owning more than one vessel was not an irrational approach. When earnings from the timber trade were good, as was the case before 1997, it was possible to recover the outlay for a new vessel in little more than one year. Since then, with the rising costs of construction timber and engines and parts, the capital outlay required for a given size of vessel has risen significantly. Nevertheless, the generally held view among vessel owners in 2002–03 was that the outlay for a new vessel of economic size, 200 cubic metres or more, could still be recovered within a period of two to four years.

Of course, to state that the capital outlay for a new vessel can still be recovered in two to four years does not mean much unless it can be set against the expected working life of the vessel. Until recently, most Madurese vessels were built from teak, with a potential working life of around 30 years. The teak-built *Bunga Bhakti*, owned by Haji Matsahri's father, was well past this age when she was lost due to shipwreck. By comparison, a vessel built of the cheapest commercial timber, meranti, will last perhaps only seven years in the humid tropical climate, long enough to recover the capital but generally considered a poor investment. Since about 1990, many transport prahus have been built in Madura of bungur and ulin, both imported from Kalimantan. Some Madurese vessels have also been built from *laban* (*vitex* sp.), the marine timber of choice in eastern Indonesia, although it is harder to find long lengths of this wood. Assuming sound construction and good maintenance, a bungur-planked vessel can last for around 25 years, while vessels planked from ulin or laban may last for as long as 40 years.[7] Given this durability of materials,

[7] The plank-to-frame fastenings on these prahus are untreated steel bolts, 16 mm in diameter. Although these rust, they can last the life of the vessel. I have seen such fastenings, exposed for plank replacement, still in serviceable condition on prahus 30 years old.

a new vessel that can recover its cost in just four years should, in principle, be an excellent investment. Yet as has been made clear, many vessels do not reach their 'natural' life expectancy.

One of the most common reasons for the premature demise of vessels, shipwreck, is now less common than it was in the days of sail, with engine power enabling more predictable courses and more accurate estimates of distance travelled for a given time.[8] However, fire remains a serious threat. Two cases of fire have been related above, those of Haji Rofi'i and Haji Matsahri's younger brother Haji Cholil. Both occurred in 2002 and in both cases the vessels concerned were almost new and laden with cargoes of timber. These were not rare instances. A leading prahu owner of Giligenting, Haji Abdul Azis, lost a vessel to fire in 1999 in Surabaya, while the remains of burnt-out vessels can be seen in most prahu harbours. Petrol-powered generators are a danger in this respect, especially coupled with the high consumption of cigarettes on board.

Most other prahu sinkings have been associated with overloading. Prahus tend to load to the absolute limit to maximise freight earnings. While such heavy loading can be safe in sheltered waters, for open-water passages it entails obvious risk. Great strains can be placed on the hulls, so that prahus which do not leak much in port can leak badly when heavily laden in a seaway. Older vessels are, not surprisingly, worse in this respect. One disturbing case encountered in 2002 involved a vessel from Sreseh. This vessel was old and in poor condition. It was also underpowered, with a very small engine and only a token sailing rig. It was carrying a cargo of timber from Marabahan, on the Barito river in South Kalimantan, bound for Pasuruan. Although the sea conditions were moderate, the vessel was very slow due to its small engine. It was also leaking badly under the weight of its heavy cargo, with a portable, engine-powered bilge pump running constantly to keep the water level down inside the hull. Two days out to sea, the pump broke down. The ingress of water was unstoppable. Bailing or pumping by hand was out of the question. As usual there was no radio, nor even a small canoe for that matter. A day later the

---

[8] Nevertheless, shipwrecks still occur. One such case happened in May 2002, when a large vessel owned by Haji Taufik of Labuhan, Sreseh (a son-in-law of the late Haji Sidik, mentioned in Chapter 4) sank after hitting a reef while under engine power en route from Beliton to Java. The cargo of poles was lost as well as the vessel but the crew were saved.

engine, by then almost under water, stopped. Desperate, the three crew members jettisoned as much timber as they could, about 30 cubic metres (out of a total of 80), to no avail. Helplessly, they waited two more days as the doomed vessel sank lower and lower in the water until it was completely awash with the three men, covered in diesel fuel, perched on top of the deckhouse. Had they been carrying salt or cement they would have almost certainly been drowned, but the cargo of timber offered a temporary reserve of buoyancy against the weight of the engine, which would otherwise have pulled the vessel under. Then on the third day, as the end seemed near, they were rescued by a small fishing boat from Sepulu. They were taken to Sepulu, where Haji Rofi'i gave them Rp 25,000 for transport and other expenses back to Sreseh.

Heavy loading also reduces the ability of a vessel to ride over the waves in the squalls that occur during the northwest monsoon. In such severe conditions the heavily laden vessels wallow and can easily be swept by waves, leading to foundering. Two such cases occurred near Madura in February 2002. One of Haji Rofi'i's vessels, an old one with a history of bad leaking, sank near the western entrance to the Madura Strait during a squall. Just one week later another Sepulu vessel, owned by a man named Haji Farouk (who had bought Haji Matsahri's *Dana*) also sank in heavy seas. In both cases the cargo was lost along with the vessel but all crew members were saved. A worse case occurred in 1999–2000 when, according to the head of the desa of Gayam, ten Sapudi vessels were lost because of the early onset of severe weather conditions.

Another common reason for the premature demise of a vessel is damage sustained in port. A prahu harbour is no place for a lightly constructed vessel. The worst port in this respect is Kali Baru, which used to be jammed almost solid with Madurese prahus. Other ports are not much better, with prahus often moored four deep against the quay. The fully laden vessels are very heavy and knocks can be punishing on the hulls. Unloading squared logs by a derrick rigged to the mast can be especially risky in this respect. As the cargo is gradually discharged, the vessel becomes lighter and lighter, heeling over more and more towards the quay as each derrick load is swung well outboard. In the final stages of unloading, the heeling can reach 30 degrees or more. While there is no danger of capsize or flooding, once the weight of the load is released, the beamy, shallow-bodied vessel can roll back with a violence that can result in serious damage if contact is made with another vessel. Many a small prahu has had its working life drastically shortened in an accident of this kind.

The lack of insurance in prahu shipping also extends to consignments. If a cargo of timber fails to arrive because of mishap to the carrying vessel, the consignor has no redress. While consignors are more likely to be worried about police checks than the possibility of the carrying vessel being wrecked, sunk or burnt, such risks are always present until the vessel reaches port.

The exception to consignments being uninsured is if they have been sent by Bulog, the government agency responsible for the distribution of rice and other essential goods. Bulog makes use of large motor prahus because they are offer good transport rates. In 1997 a large prahu owned by Haji Ariyono of Sapudi, carrying 500 tons of rice bound for Sulawesi, sank just outside Gresik harbour. Being owned by Bulog, the cargo of rice was insured. The vessel, of course, was not. The cause of the sinking was remarkable and in itself indicative of the uncertainties of prahu shipping. Coming out of the port, Haji Ariyono's vessel had to take evasive action to avoid a ship that was manoeuvering and, in the process, struck squarely against the end of the submerged mast of a Butonese prahu that had sunk at the very spot. With the great momentum of the heavily laden vessel, the mast was driven, like a lance, straight through the planking. Haji Ariyono subsequently complained to the Gresik harbourmaster, with the result that the hazard was removed. However, he received no compensation at all for the loss of his vessel. The crew members of the vessel would also have been severely affected financially because their livelihoods had been lost, at least temporarily. In prahu shipping there is no redundancy pay or compensation for crew and the owner was in no position to be able to offer them another crewing job. In the days of sail, when vessels were smaller and there were many more of them, it was much easier for an experienced sailor to find work. With so many smaller vessels being marginalised since the mid-1990s, crewing positions on good vessels have become much more difficult to obtain.

Finally, despite the many cases of vessel loss noted above, it is stressed that such cases are exceptions to the general pattern of 15 to 20 years of service before being left derelict on a beach. However, the point remains that prahu shipping is a field of business fraught with uncertainty. The high profit margins that have been enjoyed must therefore be measured against the ever-present possibility of loss of vessel and cargo. The risk of financial ruin is, moreover, greater if funds are borrowed, because just as there is no insurance on vessels or consignments, there is no option of bankruptcy as an avenue out of debt. In this traditional social milieu, debts remain until they are discharged and the moral pressure to face up to one's creditors is great. With the stakes now so high, few players can afford to remain in the business.

7

Conclusion

The dominance of the Madurese in prahu shipping along the north coast of Java from the late nineteenth century onwards was a phenomenon with its roots in strand village society. Madura's geography as an island adjacent to the mainland of Java was less important in this turn of events than was its dense population and the limited productivity of the land, forcing many Madurese to look to the sea for their livelihood. Along the south coast the orientation was naturally towards the eastern part of Java, with obvious opportunities for vessel owners in Madura; on the north coast the orientation was towards the Java Sea, just as was the case for Rembang-Lasem and the other maritime centres along the northeast coast of Java. Yet while the maritime economy of Rembang withered away in the late nineteenth century, that of the strand villages of the north coast of Madura remained vigorous throughout the entire twentieth century.

The Madurese succeeded in prahu shipping despite intense and deliberate competition from Dutch transport agencies. From the start of the twentieth century, the Madoera Stoomtram Maatschappij (MSM) developed a network of rail and bus lines, supplemented by connecting steamship services across the Strait, while during the interwar years the Koninklijke Paketvaart Maatschappij (KPM), the dominant shipping line of the Indies, did its best to undermine the economy of prahus carrying timber and mats from Banjarmasin to Surabaya, and cattle from Sumenep to Makassar.

Although Madurese prahu operators persevered in the face of this intense competition from modern land and sea transport, eventually most of the local prahu fleets around Madura succumbed, leaving only a few places from which prahus continued to be operated. The two principal export products of Madura, salt and cattle, were both important to the survival of these local fleets: salt in the case of Sreseh and Giligenting, cattle in the case of Telaga Biru and, to a lesser extent, Sapudi. Nevertheless, the importance of these two commodities for the continuation and development of these remaining fleets should not mask the significance of the cultural factors that enabled the seafarers of Madura to adapt and survive on the margins of the modern Indonesian economy. While the Madurese had long been poor, the adverse economic conditions they had endured had moulded them into a remarkably resourceful people, able to succeed where others could not. In the case of prahu shipping along the north coast of Java, the industry was already in serious decline by the late nineteenth century, but the Madurese managed to expand their operations and capture most of the market, while the mariners of Rembang-Lasem withdrew to fishing. Likewise the Madurese who migrated to inland East Java succeeded in the tobacco and coffee industries and those who migrated to West and Central Kalimantan clawed their way to economic success ahead of other ethnic groups.

The sheer economy of their operations was probably the main reason for the maritime success of the Madurese. Their vessels have always been built cheaply without compromising their seaworthiness. By comparison with vessels built in South Sulawesi, they were better able to withstand hard grounding and were on the whole considerably longer-lived. The traditional Madurese sailing rigs, in use until the 1990s, were also very much cheaper than the European-derived fore-and-aft rig used by the Bugis and placed less strain on the hulls. Crew costs on Madurese prahus were also lower, not only because of the low remuneration for individual crew members but also because Madurese prahus employed smaller crews than similarly sized Bugis vessels. The Madurese were also much slower than the Bugis to install engines in their vessels. While this slowness to modernise was partly due to lack of capital, it also reflected the economic conservatism of Madurese maritime entrepreneurs and their reluctance to spend a large amount of money on an item which, although 'fashionable', would not necessarily improve profitability. Even after engines had become standard for Madurese prahus, the emphasis on economy continued to prevail, as shown by the reliance on converted truck engines and all manner of makeshift contrivances such as plastic chemical containers for fuel tanks, with scarcely any regard for the safety regulations applicable to Pelayaran Rakyat vessels. Radios remain very much the exception and satellite navigation devices even more so.

Besides their low operating expenses, many Madurese vessel owners benefited from their involvement in trading, as opposed to carrying goods with income deriving from freight only. The Bugis had formerly been much involved as traders, but since the 1970s they fell increasingly in line with mainstream (modern) Indonesian shipping, using larger vessels to give better economies of scale in transport work. Notwithstanding their status as Rakyat vessels and the various exemptions associated with this, these large Bugis vessels are now organised more or less as small modern motor ships. In the process, however, control of the vessels has shifted away from the original prahu villages in South Sulawesi to entrepreneurs based in Makassar and Surabaya (Hughes, 1984). While some Madurese entrepreneurs, notably from Sapudi, also built very large prahus as early as the 1970s to improve their profitability from transport work, control of the vessels has remained entirely in their home villages. In general, however, most Madurese maritime entrepreneurs have still sought, where their capital allows, to profit from trading as well as carrying.

In the early years of the timber boom to West and Central Java, many Madurese vessel owners set up as timber traders, often with very little capital. Although no Madurese timber traders remain in Jakarta's main timber port of Kali Baru, Madurese involvement in timber trading remains high in Central Java and East Java, including on Madura itself. The importance of trading in the prahu economy of the Madurese is larger than it may seem to the outsider because in many cases Madurese prahus carry consignments for just one fellow Madurese timber merchant, with no other freight work.

The Madurese have also excelled in taking advantage of niche market opportunities. In earlier times this was evident from their involvement in the movement of livestock, especially horses from Sumbawa to Java and, more recently, by their domination of the pole trade and the baulk timber trade to Java. Low operating costs and flexibility have been key factors in cornering these markets, with larger and more expensive vessels unable to compete.

Finally, the profitability of the Madurese in the timber trade has depended to a large extent on their being able to operate on the margins of the state, where the relevant regulations have been weakly enforced. Other maritime groups have also benefited from weak enforcement of the timber transport regulations but the Madurese have been the most conspicuous in this respect. Since the time of the Dutch East India Company (VOC), prahu shipping has operated on the periphery of the state. In modern Indonesia, the official category of People's Shipping implies a continuing separation from mainstream shipping, with special considerations and exemptions. Although such 'special consideration' was never intended to encourage the disregard of the timber transport regulations or the evasion of timber royalties, there is no doubt that while the vessels were still small and operating under sail, many officials were sympathetic and prepared to overlook inconsistencies and shortcomings in documentation. Spasmodic efforts were made to enforce the law on the movement of timber as early as 1974 but the real push for change built up during the 1990s, coinciding with and possibly partly driven by concern for the rapidly deteriorating forest environment. Yet this tougher official stance did not greatly bother the Madurese. On the contrary, they took advantage of it by starting their own timber ports, where timber could be landed more cheaply than in Gresik or Surabaya by not paying the appropriate state revenues. Without this price advantage over timber landed in Gresik, Surabaya and even Pasuruan, it is unlikely that Sepulu and Telaga Biru would have become significant timber ports.

The boom period for Sepulu and Telaga Biru ended in 2002, when the provincial authorities finally became aware of the large volume of illegal timber passing through these ports and moved to stamp out the web of corruption associated with the traffic. Similarly, the practice of declaring only 30 or 40 per cent of the cargo carried was then no longer acceptable in Pasuruan. Since 2002 the police in East Java have been much tougher, with increasingly harsh penalties being meted out to those convicted of timber smuggling. On the other hand, there was little sign in 2003 of tighter law enforcement for the timber trade to ports in Central Java, where most of the vessels of Sumenep were discharging their cargoes.

Changing Demographics

A century ago, the main route to material success in the coastal villages of Madura was by owning a prahu. Since then the opportunities for maritime transport work have declined steadily under pressure from the development of modern shipping and land transport systems. Many local fleets of transport prahus have disappeared altogether, while among the fleets that have survived the number of active vessels is in each case much lower than during the early period of the timber boom. Nevertheless, confidence remained in the industry in Madura until about 1997, with many new vessels being built for the timber trade. Since then, however, the building of new vessels has declined sharply and many older engineless vessels have become derelict, casualties of the sluggish economy, tough law enforcement against timber smuggling and the bias towards larger vessels.

The resulting decline in employment opportunities has severely affected many households for there is little other regular work available in these coastal villages. Some men have turned to fishing but the returns from this have been discouraging because the waters around Madura have been so heavily exploited. Households that formerly relied upon income from crewing on prahus or vessel building or maintenance have been forced to seek work outside Madura. Although many from Giligenting have turned with success to the food stall business in West Java, the preferred route to prosperity is now through employment overseas, with women making an important economic contribution by working as maids in Saudi Arabia or Malaysia. The fruits of such stints of foreign work are evident in the many new ceramic-tiled houses that have been built in coastal villages in recent years. Hardly any of the new wealth from overseas work has been invested back into prahus. Instead, those with surplus capital prefer to invest in trucks or small passenger vehicles that can give a more assured financial return.

It is difficult to be certain about the number of Madurese prahus still working. During the fieldwork period, statistics did not appear to be reliable and probably included many vessels that had recently become inactive. Based on observations of all the main prahu villages of Madura during the middle of the northwest monsoon in early 2003, and taking into account the larger vessels of Sumenep that were working straight through the wet season, the total number of transport prahus still in commission then was probably less than 300, excluding the vessels of Guwaguwa and Kangean. A century earlier, in 1903, when the statistics were arguably more realistic because of the flourishing state of the Madurese prahu economy, the number of trading vessels registered in Madura was about 10 times as many, at almost 3,000. Because of the great increase in the size of vessels, the total capacity of Madurese vessels in 2003 may have been just as large as in 1903 but the total number of men employed as crew members would have been much smaller. Vessel ownership has also become more concentrated. Whereas in 1903 ownership was widely distributed with nearly as many owners as there were vessels, by 2003 many owners possessed two or more vessels.

By far the most important prahu centre in Madura has been Giligenting, where at least 100 vessels were still working in 2002–03. Sapudi, Sepulu and Sreseh each had about 50 vessels, while at Telaga Biru very few remained in commission. Looking beyond these bare facts, the largest and best equipped vessels were all from Giligenting or Sapudi, with some of these large vessels also working all year round and earning freight on both outward and inward legs. By contrast, all the vessels of Sepulu and Sreseh were earning on one voyage leg only and laying over for the northwest monsoon. There was, moreover, very little new building taking place in Sepulu or Sreseh and a distinct pessimism in both places about the future, whereas in Giligenting and Sapudi there were several large vessels under construction, indicating confidence in the upper range of the industry.

It is evident from the above that notwithstanding the fall in vessel numbers in Giligenting and Sapudi, prospects for the maritime transport industry in those places were generally better than in either Sepulu or Sreseh. In other words, the maritime entrepreneurs of Sumenep have generally been more successful than those from elsewhere in Madura. This difference deserves consideration. It is clearly not because of technical reasons or a lack of maritime capability. Nor can it be explained by the fact that most long-distance Madurese prahu shipping during the nineteenth century was from Sumenep, because prahus from other parts, especially northwest Madura, also became much involved in long-distance work during the early twentieth century.

The most likely reason for the better performance of entrepreneurs from Sumenep is the difference in external cultural influences between West and East Madura. As noted in Chapter 3, Sumenep was the cultural melting pot of Madura, with an exceptionally high proportion of 'foreign easterners': Arabs, Chinese, Malays, Bugis and Makassarese. The town itself has historically been by far the most important trading centre of Madura and its port of Kalianget was an important gateway for trade from eastern Indonesia to Java. Consequently, it was much visited by vessels from Makassar, unlike the ports in West Madura. In addition, migrant Bugis and Makassarese communities in the outer islands of Sumenep may have helped to transmit knowledge and innovations. That there was significant influence in this respect cannot be doubted for in the middle of the nineteenth century many of the trading vessels of Sumenep, and especially of its eastern islands, were types originating from South Sulawesi.

Such influence was probably strongest in the Sapudi islands, including Guwaguwa with its fleet of *padewakang*. Because Guwaguwa is so small an island, the Makassarese community there was probably dependent almost from the time of settlement on maritime trading. Later these seafarers became involved in external freight work and have continued until the present to work in the Kalimantan coasting trade. The mariners of Ra'as were also formerly very active as long-distance maritime traders and, given the proximity of Ra'as and Guwaguwa to the island of Sapudi itself, some transmission of ideas seems likely.

This notion of influence from migrant settlers in the outer islands of Sumenep may not have extended to Giligenting, where the primary orientation has always been to Kalianget and East Java. It seems likely that the success and outward-looking approach of the entrepreneurs of Giligenting owes more to the influence of another outsider group, the Chinese. The Chinese presence in Sumenep used to be strong. As a group the Chinese were much involved in maritime trade and they formed the dominant ethnic group among skippers based at Sumenep in the second half of the eighteenth century. Being based mainly in the port and town area and with connections among Chinese entrepreneurs along the north coast of Java, they were outward-looking and receptive to external ideas and little constrained by traditional notions concerning vessel types. The influence of the Chinese was especially strong around Kalianget and it is possible that they played a key role in developing the vessel-building industry at Kebun Dadap on the mainland close to Giligenting, just as they had formerly been much involved in the shipbuilding industry in north-east Java.

In West Madura, by contrast, Chinese were much less involved as maritime entrepreneurs and any such involvement probably ceased with the decline of the fleets of Arosbaya on the northwest coast and Kwanyar on the southwest. There is no evidence to suggest that the development of the fleets of Telaga Biru and Sepulu were anything other than purely indigenous Madurese affairs, while as far back as the 1870s prahu shipping from Sampang was dominated by indigenous Madurese. The prahu shipping industry in West Madura has also been seriously disadvantaged by low education standards and difficulties in raising capital. The same problems apply in East Madura but to a lesser extent. More significantly, the most successful entrepreneurs in

Giligenting and Sapudi have by their examples inspired others around them and shown the way forward. By comparison with this progressive approach, gearing up to meet the demands of the modern world, the prahu communities of West Madura have been less innovative and outward-looking.

Prospect

Although the maritime Madurese have managed well through their operations on the margins of mainstream shipping and the state, their niche is becoming increasingly untenable. Enforcement of the timber transport regulations is becoming tougher with much heavier penalties, as shown by a case in Bali in 2006 when a skipper of a motor sailing prahu from East Kalimantan was sentenced to five years imprisonment and a fine of Rp 10 billion (Rp 10,000 million) for carrying a cargo of undocumented timber (*Bali Post*, 5 December 2006). This strong stance by the state and the courts will further marginalise small operators who have depended on carrying timber with only partial or no documentation. The stricter law enforcement will inevitably favour larger vessels with their better economies of scale and which are also more likely to carry outward cargo from Java as well as being able to work throughout the year. A small number of entrepreneurs from Sapudi and Giligenting are already well established in this respect, to the extent that their vessels have been used to carry cargoes for Bulog, the state agency for the distribution of staple goods. While these successful entrepreneurs have thereby clearly moved closer to the mainstream shipping economy, their ability to keep capital and operational expenses low should continue to give them an edge over their rivals and assure their future. Nevertheless, few Madurese entrepreneurs will be able to compete in this league because of the difficulty of raising enough capital for such large vessels.

For the many owners of smaller vessels, the main problem will continue to be the cost of fuel, which since 1999 has risen steeply with the progressive removal of state subsidies for this commodity. Of course this disadvantage could be overcome, or at least minimised, by reverting to the use of sail power. The sailing rigs of smaller prahus are still extremely cost-efficient, despite their crudeness by modern yachting standards. As has been shown, some engineless Madurese prahus were still giving good returns as late as 2003. Moreover, it is not necessary to dispense with an engine altogether in order to achieve significant reduction in operational costs. A small auxiliary would still allow progress in calm and narrow waterways and, if used routinely in conjunction with full sail, could enable competitive passage times even when heavily laden and on the wind. The potential of such engine-assisted sailing for small prahus in the timber trade is not just theoretical, for two such cases were encountered in 2002–03. Both were carrying only about 50 cubic metres of timber and giving acceptable returns. As well as the benefits for increased employment opportunities, smaller vessels can go to places inaccessible to larger vessels, such as Kintap. They can also better serve the needs of small timber traders, for whom vessels of around 400 cubic metres are too large. In 2002–03, however, the only prahus making full use of sail power were already old, indicating that this technological approach is unlikely to be exploited to advantage in the future. And, of course, with each passing year the lore and skills necessary to proper use of sail power are fading away.

A vital consideration is the state of the forest resource upon which the prahu industry has long been dependent. Whereas during the 1980s there was little concern in Indonesia on this subject, there is now a deep and prevailing pessimism over the future of Kalimantan's forests. The Java Sea timber trade is perhaps the most visible factor contributing to the decline of the forest resource, although in overall terms its impact may be far less than generally supposed. Other factors, less visible but at least as serious, include the foreign export timber trade, both legal and illegal and in both roundwood and sawnwood form; the plywood industry, again mainly for the foreign export market; the consumption of non-plantation forest wood by wood pulp and rayon factories; and, potentially the most serious of all, forest clearance for oil-palm plantation agriculture. In the case of timber exported to Java, the benefits, in a material and durable form, have at least flowed through to a large number of people, while also allowing some hydrological relief for the heavily exploited forests of Java.

Nevertheless, the inescapable fact is that present logging and land clearing rates are unsustainable. West Kalimantan is the worst affected province, but other provinces are also badly affected. The case of the timber industry at Sungai Kaki, as detailed in Chapter 5, is a grim harbinger of other such cases to come. According to the Director of Economic Planning for the kabupaten of Kotawaringin Timur in Central Kalimantan in 2003, the district government could expect to rely for its income on the timber industry for only a few more years. Like so many others, he saw the economic future as depending on the oil palm industry.

Yet despite such gloomy forecasts, the timber industry will not suddenly cease to operate. Rather, it will gradually slow down, with vessels taking longer to obtain full cargoes and the timber being of poorer quality. The lower quality may make a difference for the export trade but should not matter much for domestic consumption. However, the dwindling nature of the forest resource highlights the urgent need for larger prahus to diversify away from total dependence on the timber trade.

Finally, it remains to be seen how the maritime Madurese will be affected by the completion of the Suramadu bridge. The benefits of this road link to Surabaya, opened in 2009, will be most apparent in those parts of Madura closest to the bridge, especially Sukolilo and Bangkalan. These places are primed for development, with a trickle-down effect throughout the rest of the island of Madura. Local populations in places such as Kwanyar, which a century ago was already losing its position as a seafaring centre because of competition from land transport, will experience increased economic opportunities. As far as the prahu transport industry is concerned, however, the opposite would normally apply. If the bridge had been built a century earlier, the prahu fleets of the south coast of Madura would have been marginalised more rapidly than actually happened. Now, it is possible that the new land link to Surabaya may be a boon for the maritime entrepreneurs of Sepulu, recently so hard pressed. With Sepulu being at least as close as Pasuruan to Surabaya, yet a day closer to Kalimantan, the Sepulu timber merchants should have a significant market advantage over their rivals with benefits flowing on to the local maritime transport industry and the community as a whole. For this to be effective, however, timber smuggling will need to be eradicated at Pasuruan as well as in the ports of Central Java. If this can be achieved, Sepulu will be poised to reap the fruits of the development boom taking place in nearby Bangkalan and might become once more an economic inspiration for the rest of Madura.

Whatever the impact of the new bridge, the greatest challenge now facing the maritime entrepreneurs of Madura is that of modernity. In a general sense, it has always been so. As we have seen, the Madurese have shown remarkable ingenuity in adapting to a changing world, making the most of limited opportunities while retaining their 'traditional' ways of doing things. Thus through the first half of the twentieth century, Madurese sailing craft withstood the modernising forces of the late colonial state and the commercial pressures from the KPM and the Steamtram Company, until Independence and the weakening of regulatory controls brought new opportunities. Then through the three decades of the New Order period, when the state gradually modernised its sea communications, the timber boom allowed the seafaring Madurese to carry on business in their own way and with little regard to maritime or forestry regulations. This strategy is at last no longer viable. Although the category of People's Shipping still exists as an officially recognised informal sector, it has become too risky to attempt to gain a competitive edge by disregarding the state's laws and regulations. If the seafarers of Madura are to survive as shipping operators into the twenty-first century, they will need to modernise not just their maritime technology but also their approach to business. The replacement of engines has already caused a sudden and dramatic change in maritime culture. How business and social culture will adapt to the breakdown of geographic and cultural isolation remains to be seen.

Appendix 1

Origins of the *golekan*, *janggolan* and *lete-lete*

The three traditional vessels used for the greater part of the twentieth century by the seafarers of Madura, the *golekan*, *janggolan* and *lete-lete*, have been described in general terms in Chapter 3. This appendix provides a more detailed explanation of their development and lineages.

These three Madurese prahu types stand apart from other contemporary traditional types. Whereas the vessels of Sulawesi and eastern Indonesia were built by dowelling together short pieces of wood hewn to the desired shape, with no bending involved, the Madurese builders used long planks, bending them over fire before fixing them in place with dowels. This was the same technique used by the Javanese but otherwise there was little similarity with Javanese vessels. The basic Javanese hull type, as represented by the *mayang*, had a multi-chined form with distinct angles in cross-sectional shape rather than the rounded hulls of the Madurese, and the hull was reinforced by a series of bulkheads. The Madurese did adopt this type of construction for some of their fishing vessels but not for their main transport craft.

In contrast to the broad and shallow *mayang*, the basic vessel of Madura was the *peduang* (*bedouang*, *paduwang*), a double-outrigger canoe. The largest versions of these had planked hulls and a rather 'boat-like' form by comparison with the pure dugout canoe and a cargo capacity of several tons. This outrigger ancestry is most obvious in the case of the trading prahu of northwest Madura, the *golekan*. Not only was the hull shape of the *golekan* more canoe-like and less capacious than most other types but also, and more significantly, the sailing rig was directly derived from that of the *peduang*.

The *peduang* had two broad triangular sails. The foresail was slung from a short mast stepped in the bow but the much larger mainsail was supported differently. In the original form, there was no mainmast at all. Instead the lower end of the long bamboo yard was secured to the deck, while the upper end of the yard was supported by a prop that was held upright by rope stays attached to the outrigger beams (see Figure 2, p. 76). Later, as *peduang* became larger, a short mainmast was added to make the rig easier to manage (Figure 3).

This same arrangement was used by the builders of the *golekan*, except that because there were no permanent outriggers, a stout beam (*tanjoran*) was extended well outboard on the windward side to provide the same low angle for the stays supporting the upper ends of the yards. Two beams were used, one for each side, with the beam not in use slid inboard so that it would not drag in the water (Figure 4). In port, both beams were brought inboard. As *golekan* became bigger, a prop was used on the upper end of the foresail yard as well as on the mainsail yard, both being stayed to the one extended beam.

Trading *peduang* were common around Madura during the nineteenth century but disappeared in the early twentieth century, apparently not long after the emergence of the *golekan*. The shift away from outriggers was a natural development as the builders sought to design more capacious hulls. Also, outriggers were very inconvenient in crowded ports. Nevertheless, the popularity of the *peduang* until the modern era was not because the Madurese were unable to build beamier boats that did not need outriggers. Since at least the fourteenth century, the Madurese had been well known for their skill in building and using boats (Scott, 1982: 340). These boats were generally leaner and more canoe-like than the Javanese *mayang* and of different construction. They were open boats, propelled by oars as well as sails, and were used for tasks for which outrigger craft were unsuitable, such as local transport or net fishing.

On the north coast of Madura, these small open boats came in two types, distinguished by the construction of the ends. One type, probably the older, had no stem and sternpost. Instead, the keel extended a short distance beyond the hull proper at each end, while the planking above terminated on narrow transoms, a type of construction known as 'bifid-ended' (c.f. Hornell, 1970). This type of boat was called *alis-alis* (often shortened to *lis-alis*), literally meaning 'eyebrows'. This name may have been given because of an ornamental motif resembling a pair of eyebrows that was often applied to the

Figure 3 An outrigger trader in the Madura Strait near Surabaya in 1856, from a painting by Thomas Baines, reproduced in Braddon (1986: 110). Like Paris, Baines was an experienced seaman and a careful observer of small craft.

Figure 4 A trading *golekan* at Singapore, about 1925, from a photograph taken by W.M. Blake and reproduced in Hawkins (1982: 74).

top of the bow. In any case, eyebrows are an important aspect of feminine beauty in Indonesian culture and these boats were strongly associated with the feminine, so much so that they were commonly referred to in northwest Madura as *parao bini* (female boat). *Alis-alis* were never coarse of build and the best were exceptionally graceful craft and decorated or 'made up' like a woman to look her finest.

The other type of small open boat of northwest Madura had conventional ends, with stem and sternpost. These boats were called *golekan* and it was from this type that the fully decked seagoing *golekan* was developed. Whereas the *alis-alis* was 'female', the *golekan* was 'male', so that it was widely referred to as *parao laki* (male boat). These boats were also given much ornamentation but the motifs used were different from those on the *alis-alis*. The primary motif on the *golekan* was the rooster, associated in Indonesian culture with combat and fearlessness, while the overall effect was one of austere handsomeness, in keeping with the cultural ideal of the male.

The seagoing *golekan* was a purely indigenous vessel in all respects. This was evident not only from the hull form and styling and the sailing rig derived from the *peduang*, but also from the construction. The planking was very thick and braced by a series of thwarts arranged in tiers and notched through the shell (see Hawkins, 1982: 63). This archaic method of reinforcement, not recorded for transport prahus of Java, was developed from the system used in the inshore boats of the Madurese.[1]

---

[1] The construction detail drawing from Blake (1925), reproduced in Hawkins (1982: 63), shows the planking of the *golekan* as being 4½ inches (115 mm) thick, far heavier than in modern prahus of comparable size, including in northwest Madura. Was Blake mistaken in this respect? Because his work was based on observations of working *golekan* calling at Singapore, he may not have had the opportunity to verify the plank thickness empirically. However, as improbable as 4½ inch thick planking seems, it is supported by my observations of the building of traditional fishing boats at Kwanyar on the south coast of Bangkalan in 1984. The planking on these open boats was up to three full inches (75 mm) thick for a hull only nine metres long and of only moderate beam and depth. The planks were also long and broad, keeping joints and seams to a minimum. The bracing of the hull was light, consisting of a few thwarts notched through the planking. Kwanyar was formerly an important centre for seagoing prahus. It seems that elsewhere in Madura, planking stock had become lighter as transverse framing was added, a result of Javanese, Chinese and European influence.

Photo 45 Small *alis-alis* used for fishing, Telaga Biru, 1986. In very small craft like this, the outrigger arrangement to extend the staying base for the rig consists of a light pole extending outboard on both sides, secured to the hull with rope lashings and wooden levers.

Concerning the etymology of the name *golekan*, the Dutch observer van Deventer (1904: 110) wrote that it meant 'freight seeker', adding that this was "an indigenous word which conveys more than a whole treatise". That might have been so if van Deventer's meaning had been correct, which it was not. A different explanation was offered by Gibson-Hill, who stated that the name was derived from the Malay word *golek*, which he claimed meant 'crank' (tender, or lacking initial stability). He reasoned that this name had been given to these Madurese vessels because in the early stages of development, when the hulls were presumably narrower and more canoe-like than they later became, they were rather unsteady in the water (Gibson-Hill, 1950: 126). There is a semblance of plausibility here but the argument is wrong. If this etymology were correct, why was the same name not given to the equally crank *alis-alis*, from the same locality?

The true explanation is simple enough. The name is derived from the word *kolek*, a Javanese word for 'boat', applied to a wide range of small craft along the north coast of Java. It is also the name for the oldest and most widespread type of vessel in the *mayang* family, a type large enough for cargo work and which predated the *golekan* of Madura by at least several centuries. The term *golekan*, '[a kind of] *kolek*', was used for the conventionally ended 'male' boat because it had a distinct stem and sternpost like the *kolek* of Java. This name was not given to the 'female' boat for the reason that having bifid ends, it differed markedly from vessels of the *kolek* type.

It seems unlikely that the preference for the male prahu (the *golekan*) for long-distance work was based on superior seaworthiness of the hull form. While the *golekan* may have been popularly regarded as more seaworthy, large *alis-alis* would have been just as capable. Indeed, at Tamberu on the central north coast, vessels of the latter type (called *jabar*) were used until as late as the 1970s for voyages to Kalimantan. One knowledgeable informant there, from an extended family of sailors and boatbuilders, claimed that the 'female' hull form was the more capable at sea because it was drier forward. Possibly the 'male' type became the choice for long-distance trade because of cultural associations of such voyaging with the masculine ideal. On the northwest coast small *alis-alis* remained popular until recently for inshore fishing, although since about 1990 no more of the type have been built there.

The vessel type of the southwest coast, the bifid-ended *janggolan*, was related in both structure and form to the *golekan*. Nevertheless, its course of development was quite different. Along the south coast of West Madura the bifid-ended type was used almost exclusively, except for small canoes. Indeed, until as late as 2003 the *alis-alis* remained overwhelmingly popular as a fishing craft at Sukolilo on the south coast of Bangkalan, while a larger version, the *kroman*, had been used in the same area for more than a century for inshore transport work. *Alis-alis* are no longer found east of Sukolilo, but up until the 1970s they were formerly the small craft of choice as far along the coast as the Blega estuary. Small bifid-ended boats were also common on the Java side of the Madura Strait, possibly as a result of Madurese migration. The bifid-ended style is not recorded for the rest of the north coast of Java, the realm of the *kolek* and the other members of the *mayang* family.

The sailing rig of the *janggolan* was a more highly developed version of that used on the *golekan*. Because of the large size of these vessels, the heavy outrigger beams (*antol*, c.f. *tanjoran* on the north

Photo 46 A *janggolan* moored at Kali Baru, 1984. The outrigger beams are much heavier than those used on *golekan* and are usually left extended in port except when lying alongside for loading or unloading.

coast) were both left in the extended position at sea and also often at anchor, being brought inboard only for berthing in crowded harbours (Photo 46). Similarly, the yards (*pengba'un*) and their props (*sokongan*) typically remained in position for the entire sailing season.

Despite the sailing rig of the *janggolan* being clearly derived from outrigger craft, there is no living memory, even of stories handed down, of outrigger transport vessels at either Sreseh or Sukolilo. However, in both places I interviewed elderly men who had a definite and single opinion on the origin of the *janggolan*. These men were all emphatic that the provenance of the type was not Madura at all but a place in East Java called Jatisari. No one knew exactly where Jatisari was, but it was thought to be on the lower Solo River. There is in fact a place called Jatisari on the lower Solo, about 30 kilometres from the sea in the kecamatan of Glagah. In 1999 there were motorised riverboats there, with bifid ends. They had no sailing rigs, were much narrower and less capacious than *janggolan*, and were roughly built. Nevertheless, the ends bore the same generic ornamental motifs as found on both *alis-alis* and *janggolan*, although they were more crudely done.

Could this place be the Jatisari of *janggolan* legend? The river-boats observed give at best limited support for the idea. However, although the lower Solo is today a backwater, before the Dutch diverted its mouth away from Gresik in the nineteenth century it was the main transport artery from the coast to the interior of Java, plied by numerous vessels (Dick and Rimmer, 2003: 119–29; also Schrieke, 1957: 112–17). These vessels were quite large but relatively narrow of beam, as befitting river craft. They had no fixed sailing rig and were moved against the current by means of oars, poles and ropes. Nevertheless they did have bifid ends "with the fore and after posts projecting high and outward beyond the hull proper" (de Bruyn Kops, 1854: 61).

Vessels of this same type were also used in the mid-nineteenth century as sailing lighters in the Surabaya roads, adjacent to Gresik harbour, carrying goods back and forth between ships and the shore. These were also narrow craft, with a beam of only 8 or 9 feet on a length of 45 to 50 feet, which together with the other evidence suggests a riverine origin. Fortunately, the sailing rig of these lighters was carefully described. They had two very short masts, one in the very bow and the other, a third of the way aft, as well as long bamboo yards with the main yard supported at its upper end by a pole which in strong breezes was stayed to the end of a temporary outrigger beam (de Bruyn Kops, 1854: 64). This was exactly the same sailing rig as was later used on the *janggolan*. There can be no doubt that these sailing lighters were the predecessors of the *janggolan* and de Bruyn Kops' careful description strongly supports the belief that these vessels originated in the lower Solo River.

The *janggolan* was thus not a direct development from small inshore boats as had occurred on the north coast, but rather was developed from 'bulk carrier' riverine prahus that became used as sailing lighters at Surabaya and Gresik, with the addition of the rig derived from the *peduang*. To judge from a drawing by Paris (1842: plate 89) of a transport boat in the Surabaya roads in the 1830s, Madurese watermen were then involved in ship-to-shore work but using open boats powered by sail and oars. In due course the *janggolan* type was copied by builders on the Madura shore, who made their own modifications while retaining the essence of the design.

While other large vessel types might have served just as well, for the people of the strand villages of southwest Madura the *janggolan* exerted a special appeal because of the elaborate ornamentation of the ends. This decoration of the vessels provided an avenue for artistic expression but there was also a deeper cultural symbolism involved. Like the *alis-alis*, the *janggolan* was regarded as 'female' and the ornamental motifs used were mostly connected with the feminine. Floral motifs predominated and 'flower' names were very popular for the vessels. The motifs were arranged in a specific pattern with considerable scope for variation in the finished forms so long as their essential character was unchanged (Photo 47). Although the *janggolan* was developed only in the nineteenth century, the principal elements of the ornamentation can be traced deep into the Southeast Asian past and examples of both their use on boats and their association with the feminine exist from the Sulu Sea to the Malay peninsula.[2]

Why were these vessels called *janggolan*? In a brief description of the vessel type, the *Encyclopaedie van Nederlandsche-Indië* (1917, vol. V: 432) notes that the name literally means 'transport' or 'carrier-vessel'. However, Burningham and Mellefont, observers with extensive experience of Indonesian watercraft, cast doubt on this meaning. They suggest instead that the name may be derived from *jang*, a particle word standing for 'plank' or 'strake', coupled with *olan* as a possible

[2] The most significant evidence of this comes from two photographs in Feldman (1985) of male and female Bajau gravemarkers from the Sulu area of the southern Philippines. Both gravemarkers are in the form of beautifully carved miniature boats with upright structures mounted amidships. The male boat has a form like a *golekan* and is topped by a decorated post, representing a penis, whereas the female boat is bifid-ended like the 'female' boats of Madura and is topped by a slab carved with motifs strikingly similar to some found on the *janggolan*. This same association with mortuary art also exists in Madura, where female gravemarkers are in the form of upright slabs, often tapered towards the bottom, whereas male gravemarkers are in the form of posts. Further, the female gravemarkers in Madura are often decorated with a motif called *pucuk repung* (bamboo shoot), which is also the principal motif on the ends of the *janggolan*. This motif corresponds to the *pako rabung* of the southern Philippines, referred to by Baradas (1968: 139) as "the gravemarker of the female sex". For more examples of male and female gravemarkers of these generic forms, see Kiefer and Sather (1970: Plates 1, 2). I have also observed similarly shaped gravemarkers in West Malaysia, as well as stylised versions of the *pako rabung* on traditional fishing boats of Kelantan.

Photo 47 The bow of the *janggolan*, *Si Sekar Mendung* (Solemn Flower), built by the late Haji Umar of Batuputih, photographed at Sampang in 1983. Although the upper motifs in this case are of Dutch origin, the principal motifs are deeply traditional and widespread across island Southeast Asia.

cognate for *ulang*, 'repeat': "thus a *janggolan* is a vessel with (many) more strakes than a *lis alis*" (Burningham and Mellefont, 1997: 36). This argument is ingenious but unconvincing on linguistic grounds: the resultant word would be *jangolan*, pronounced with no hard [g], unlike *janggolan*. Burningham and Mellefont were nevertheless justified in being skeptical of the claimed meaning of 'transport', for the name *janggolan* is not known to have been used for any of the other transport vessels of Java and Madura, nor is there any word *janggol* in Javanese or Madurese meaning 'to carry' or similar. No-one in Sreseh or Sukolilo can explain the origin of the word. For the mariners of Sreseh, the term *parao janggol* means a vessel that is used for extended voyages to distant places. This formerly contrasted with *parao pegon*, applied to vessels working in local waters, including to the far side of the Madura Strait. The name *pegon* was eventually dropped but is still remembered by many elderly people.

The key to this conundrum lies, unexpectedly, in Bali. The main port there, Benoa, is situated on a broad natural harbour with a fishing village, Tanjung Benoa, at the tip of an isthmus on the far side of the harbour. Until at least very recently, the most common form of transport across the harbour was by small open ferry boats called *janggolan*. It is hardly likely that this name derived from any link with southwest Madura and these ferry boats bore no resemblance at all to the *janggolan* of Madura. Indeed, the hull shape was not indigenous at all, being much like that of a nineteenth-century European ship's boat, including the central rudder. Benoa is not one of the main trading ports of the archipelago as it is inconveniently located. These days it is used mainly by fishing vessels, pleasure craft and the occasional tourist ship. In the middle of the nineteenth century, however, it became a bustling place after a Danish trader, Mads Lange, opened a very successful entrepot business at nearby Kuta (Nordholt, 1981: 29-30). In the heyday of this entrepot, numerous European ships called at Kuta, so that there was great need of small vessels for lightering and general ship-to-shore work. Taking into account the origins of the *janggolan* of Madura as a sailing lighter in the ports of Surabaya and Gresik, it is very likely that the use of the same name for the small ferries of Benoa harbour is a legacy of the period of the Kuta entrepot. We may therefore conclude that the word *janggolan* was originally a widespread colloquial term in Central Indonesia for a transport vessel but of a particular type, namely one engaged in some sort of lighterage service between ship and shore. It was apparently also used sometimes for local passenger services on land in Java but has since fallen into disuse.[3]

The development of the third major traditional type, the *lete-lete*, is more obscure. Although a wide range of prahu types existed in Sumenep waters during the mid-nineteenth century, the *lete-lete* was not among them. The main trading prahus included vessels with European influence such as the *toop*; vessel types which had originated from South Sulawesi such as the *pajala*, *sopet* and *padewakang*; and several types of vessels showing Javanese and Malay influence, notably the *mayang* and the *pencalang*. The last was a Malay trading vessel that had found strong favour along the north coast of Java and as far away as Makassar. The Javanese examples of this type were called *cemplon*, while the Madurese developed their own version, the *kacik*. There was also a local version of the *golekan* used on the north coast of Sumenep and in the Sapudi islands, similar in form to the vessel of the same name from West Madura but smaller and with more modern construction and simpler rig. None of these vessels showed any direct development from outrigger craft, although there were plenty of outrigger craft in East Madura. It is probable that the change to more capacious vessels without outriggers had not been an 'evolutionary' process but rather one of replacement of outrigger vessels by unrelated vessel types.

In the first or second decade of the twentieth century, this great diversity of prahu types in East Madura changed after the emergence of the *lete-lete*. As this new vessel gained rapidly in popularity, the vessel types from Sulawesi disappeared from Sumenep waters, while the *kacik* also declined in popularity against the *lete-lete* until it too disappeared.

The *lete-lete* was not a direct development from the *golekan* of the north west coast, as has been assumed by some observers. Technically it could be described as a *golekan* because it had conventional

---

[3] Pers. comm., Professor A.M. Djuliati Suroyo, Diponegoro University, Semarang. According to a Madurese sailor, the word *janggol* is often used to refer to harbour prostitutes in Sundanese-speaking areas such as Cirebon. Such usage could overlap with the notion of a 'shuttle service', especially in the form of the small harbour transport boats such women use. Another informant from Sreseh stated that her late father was convinced that the word had originated from Cirebon, suggesting that he had heard it used by locals there. Even so, its usage was clearly more widespread.

Photo 48 A *golekan* of the north coast of Sumenep, 1986. Although superficially similar to the vessels of the same name from the western half of the north coast, the construction is quite different. They are also smaller and less wide-ranging in their activities.

ends, like the boats of Java, and this name was sometimes formerly used in Sumenep for the *lete-lete*. Nevertheless, it was a very different vessel from the true *golekan*. First, the sailing rig was completely different, being essentially a freestanding single-masted rig, without the complex staying arrangement of the vessels of West Madura. The mainmast was taller and much stouter, while the main yard was also much stiffer, being a 'laminated' spar of two or more large bamboo poles strongly lashed together. The methods of sail handling, including tacking, were also quite different from those used in the trading vessels of West Madura. Second, the hull was quite different from that of the northwest *golekan*. The shape was broader and shallower and the shell was reinforced with heavy grown frames, as in the European tradition, rather than the archaic system of thwarts notched through the hull. Perhaps most revealingly, the plank structure was different, indicating a different origin. In particular, the shape of the garboard (the first strake, or run of planking, from the keel, which influences the subsequent plank pattern) was different from that found on the vessels of West Madura.[4] On the other hand, the aesthetic detailing of the *lete-lete* was similar to that of the *golekan*, particularly in the common use of a swirling motif, the *okelan*, on the stem and sternposts, and a small swept-up finial just inside the stem and sternpost corresponding with the larger scroll-shaped finials on the *golekan*.

Before the emergence of the *lete-lete*, the dominant type of trading vessel around the Java Sea had been the *pencalang* (or *cemplon*, or *kacik* in Madura). This vessel had proven itself over the two previous centuries as both a load carrier and sea boat and was deservedly popular with the Bugis and Makassarese, as well as with the Javanese and Madurese. However, while the *pencalang* was an excellent vessel for trading purposes, it was not easy to build for men coming from the maritime traditions of Madura or Makassar because the forefoot (the lower stempost and adjacent plank structure) was deeper and more angular than in the traditional vessels of these two areas.[5] The *kacik* also had much aesthetic embellishment, especially in the form of ornamental carvings that were considered integral features of the type, requiring specialist craftsmen.

For the vessel owners of the south coast of Sumenep with their fondness of decorative carving, the complication and expense involved in obtaining the services of such specialist craftsmen was acceptable. However, in the Sapudi islands, where the *lete-lete* was first developed, aesthetics were a luxury tempered by economic necessity. The Sapudi group, including the islands of Ra'as and Guwaguwa, was also further removed from the influence of the high culture of Sumenep and, at the same time, closer to the communities of Bugis and Makassarese who had settled in the outer islands. Ornamentation was never as important in the maritime culture of South Sulawesi as it was in Madura.

The new type of vessel developed in the Sapudi islands was essentially a melding of ideas taken from the South Sulawesi types and the Malay-Javanese *pencalang*, grafted on to the hull of the *golekan*

---

[4] In the early 1980s, when there was a strong demand in Sreseh for the building of *janggolan*, two boatbuilders were hired from Giligenting to help with this work. These two men had never built a *janggolan* before, but no problems were anticipated for they were both accomplished craftsmen. However, to everyone's surprise, after two weeks they gave up and returned to Giligenting because of the difficulties they encountered in fitting the garboard on a new vessel.

[5] The Bugis were formerly unable to build the *lambo*, which had a very deep forefoot. When they did eventually switch to this type for their smaller vessels, the forefoot was less pronounced than in the Butonese model.

of Sumenep (not that of West Madura). The primary inspiration for the hull shape was undoubtedly from South Sulawesi. Gibson-Hill (1950: 127) noted that the *lete-lete* hull bore a considerable resemblance to that of the *pajala*, the basic vessel of the Makassar coast. This was an astute observation, the more so because Gibson-Hill was probably unaware of the many vessels of South Sulawesi origin in Sumenep waters during the nineteenth century, as noted by Bruyn Kops (1854). Rather than the *pajala*, however, it is possible that the most direct influence in this respect came from another type of Sulawesi vessel common in Sumenep waters, the *sopet*. This had the same basic form as the *pajala* but had a single heavy mast stepped on the keel in the same manner and position as on the *lete-lete*, whereas the *pajala* employed the deckstepped tripod mast arrangement standard among vessels of South Sulawesi. Indeed, until the 1980s there were many small vessels in eastern Sumenep looking much like small *lete-lete* but referred to locally as *sopet*.[6]

While the general form of the *lete-lete* was inspired by the basic vessel model of South Sulawesi, there was also clear evidence of influence from the *kacik*, notably in the form of the low open bulwarks formed by all the frames in the main part of the vessel being extended above the deck (Photo 49). This was a distinctive feature of the *pencalang* family of vessels but not one associated with the traditional vessels of Sulawesi, nor with the *golekan* of West Madura. Another feature borrowed from the *kacik* was the small square-shaped 'doghouse' at the after end of the gabled main deckhouse: though not used on the Sapudi vessels, this 'doghouse' became standard on *lete-lete* from elsewhere. The sailing rig took the sturdy mast arrangement of the *sopet*, but with a triangular sail as used on the *golekan* and other small local boats instead of the standard rectangular sail of South Sulawesi. The rudder support system was also clearly Madurese, with a single lateral rudder slung on a central post rather than the more complex gantry system used in South Sulawesi. Finally, the aesthetic embellishment, with the swept up finials at each end and the *okelan* motif, obviously derived from the *golekan*. However, rather than being a 'borrowed' feature, this distinctive styling was inherent in the new vessel. In other words, despite the considerable external influence upon the development of the *lete-lete*, it was a true Madurese

[6] The same name was also used for smaller vessels of unrelated type used by Bajau people who had settled in the Kangean islands.

Photo 49 The low bulwark on a *lete-lete* from Sapudi.

creation. After a long period of domination by 'foreign' prahu types, the indigenous maritime culture of Sumenep was here re-asserting itself through the development of a vessel broadly based on a pre-existing local model but incorporating features from other sources to create a completely new but indisputably Madurese craft.

The *lete-lete* quickly became the dominant trading vessel of Sumenep for three reasons. First, because of the styling and ornamentation of the ends, it retained much of the intrinsic appeal of the *golekan* and was regarded as being 'Madurese'. Second, it was much cheaper to build than the *kacik*, while retaining the desirable features of the latter as a trading vessel. Third, it was an excellent sailer. Indeed, such were its qualities in this respect that the type spread to the Makassar and Mandar coasts of South Sulawesi, even though the ornamental aspects identifying it as Madurese were there discarded completely. According to a Mandar informant interviewed in the 1930s by the maritime ethnologist C. Nooteboom, the *lete-lete* was better able to stand up against a strong breeze and could point higher into the wind than the existing South Sulawesi types (Nooteboom, 1940: 30). Nevertheless, the *lete-lete* did not become dominant in South Sulawesi because it was more awkward to tack than the European-influenced rigs already well established in those waters. For this same reason, the *lete-lete* finally lost favour even in Giligenting, with most being converted to sloop rig after 1990.

The origin of the term *lete-lete* is unknown. Possibly it was a nickname given to distinguish this new sort of *golekan* from the more traditional vessel of the same name. In the early twentieth century the type was referred to in some publications as *tekletek*. However, this appears to be the same name, with the first syllable dropped: [*le*]*tek-letek*, as in [*a*]*lis-alis*. Such fore-shortening of duplicated names is a common feature of Madurese. The [k] as written stands for a glottal stop, as in *nasik*, a spelling formerly encountered on menu lists of some small eating establishments, for *nasi* (rice). Piollet (1995: 82) notes that *lete-lete* were sometimes referred to as *magonan*, a name possibly derived from the word *pegon*, formerly used in Sampang-Sreseh to denote a small to medium transport vessel for use in the Madura Strait. The name *pakesan* was also often used on the south coast of Pamekasan and Sumenep, with reference to the *pakes*, the name in East Madura for the finial structure just inside the stem and sternpost.

Appendix 2

The timber transport permit

Nomor Seri : DC 0689644

DEPARTEMEN KEHUTANAN
DIREKTORAT JENDERAL BINA PRODUKSI KEHUTANAN

PEMERINTAH KABUPATEN TANAH LAUT DINAS KEHUTANAN

SURAT KETERANGAN SAHNYA HASIL HUTAN (SKSHH)

UNTUK *) Kayu Olahan

Berlaku Lima hari, dari tanggal 19-11-2002 s/d 23-11-2002

Dinas Kehutanan Propinsi : Kalimantan Selatan.

Dinas Kehutanan Kota/Kabupaten/CDK : Tanah Laut.

Asal Hasil Hutan :

- Nama dan Alamat Lengkap Kantor Perusahaan Pengirim / Penjual / Perorangan: UD.Bahagia Setia Bakti-Usaha Ds. Nusa Indah. Kec. Bati-Bati. Kab. Tanah Laut.

Tujuan Pengangkutan :

- Nama dan Alamat Lengkap Kantor Perusahaan Pembeli / Perorangan: PT.Kayan Jaya Tanjung. Jl.Margomulyo Indah B.58 Surabaya Jawa Timur.

- Pengangkutan melalui : Darat/Laut/Sungai/Udara **)
- Jenis Alat Angkut : PL. Nor Buat.
- No. Polisi / Nama Kapal / No. Lambung / Nama Tongkang :

- Alamat lengkap Pembeli / Penerima / Lokasi Pengolahan / Perorangan : Surabaya Jawa Timur.

- Alamat Lengkap Lokasi Tempat Muat: Pelabuhan Muara Kintap. Kec. Kintap.
- Alamat Lengkap Lokasi Tempat Bongkar: Pelabuhan Surabaya. Jawa Timur.

NO URUT	NOMOR DHH	JENIS HASIL HUTAN	JUMLAH (BTG/LBR/KPG/BDL)	VOLUME/BERAT (M3 SM/TON)	KETERANGAN
1	2	3	4	5	6
1.	16/DHH-BSBU/ XI/2002.	Terantang. Mahang.	=====500==== ===1.999====	===18,0864=== ====11,9940==	PSDH dan DR Telah Dibayar.
JUMLAH:		DENGAN ANGKA *)	===2.499====	===30,0804===	
		DENGAN HURUF ***)	Dua Ribu Empat Ratus Sembilan Puluh Sembilan Potong Sama Dengan Tiga Puluh Koma Nol Delapan Nol Empat Meter Kubik.		

Lembar ke - 1 dan ke - 2
Dokumen ini diterima oleh Petugas Kehutanan
Pada tanggal .....
Nama : .....
Jabatan : .....

Hasil Hutan telah diterima
oleh : .....
di .....
Pada tanggal .....
Nama : .....
Jabatan : .....

Pelaihari, 19 Nopember 200[illegible]
Penerbit SKSHH,
Nama Rasam.
Jabatan Penerbit SKSHH KO HR.
No. Reg. 048/BC/TALA/KO/HR/Rsm

MELENGKAPI BERSAMA

Penjelasan :

1. Dokumen ini merupakan Surat Keterangan Sahnya Hasil Hutan sesuai Undang-Undang Republik Indonesia Nomor 41 Tahun 1999.
2. Dokumen ini wajib dilaporkan kepada Petugas Kehutanan yang ditunjuk selambat-lambatnya 1 (satu) hari setelah hasil hutan sampai di tempat tujuan.
3. Hasil Hutan yang dilengkapi dokumen ini sah apabila isi yang tercantum dalam dokumen sama atau telah sesuai dengan fisik hasil hutan yang diangkut dan tidak ada perbedaan isi/volume lembar ke-1 s/d ke-6 serta tidak terdapat perubahan yang berupa coretan, tindasan atau hapusan.
4. Barang siapa menyalahgunakan dokumen ini diancam dengan sanksi sesuai Undang-undang Nomor 41 Tahun 1999.
5. Barang siapa meniru, memalsukan dokumen dan atau dengan sengaja menyimpan serta menggunakan dokumen tiruan atau dokumen palsu diancam dengan hukuman sesuai peraturan perundang-undangan yang berlaku.

Catatan :

*) Diisi : KB / KBK / LIMBAH / BBS / HHBK , sesuai dengan hasil hutan yang dilengkapi dokumen ini.
**) Coret yang tidak sesuai.
***) Diisi dengan huruf mengenai total dari jumlah Batang/Lembar/Keping/Bundel/ dan Volume/Berat (m3:sm:ton).

Model : DK. B. 101

Lembar ke - 1

DAFTAR HASIL HUTAN (DHH)
UNTUK KAYU OLAHAN

Nomor : 16 / DHH - BSBU / XI / 2002 Tanggal : 18 Nopember 2002

CDK/Dsihut.Kab/Kodya : Tanah Laut Nama Perusahaan : UD. Bahagia Setia Bakti Usaha
Dinas Kehutanan Propinsi : Kalimantan Selatan Alamat Perusahaan : Desa Nusa Indah Kec. Bati, Bati

Nomor	Kelompok Kayu Olahan	Jenis Kayu Olahan	Ukuran (cm)			Jumlah Keping	Volume Kayu (M3)
			T	L	P		
1	Rimba Campuran	Terantang	6	20	400	127	6,0960
			8	12	400	130	4,9920
			6	12	400	243	6,9084
	Jumlah					500	18,0864
2	Rimba Campuran	Mahang	4	6	400	833	7,9968
			3	5	400	333	1,9980
			2	3	400	833	1,9992
	Jumlah					1.999	11,9940
	Jumlah Total					2.499 Ptg	30,0804 M³

DHH. Ini merupakan lampiran SKSHH Nomor Seri DC.06896441 Tanggal 19. Nopember 2002

Diperiksa Oleh
Pejabat Penerbit

Ram.
No.Reg.048/BC/TALA/KO/HR/Rsm.

Pelaihari, 18 Nopember 2002
Dibuat Oleh :
Perusahaan / Pemilik

HAJI SULAIMAN

The above two pages are a copy of an SKSHH (certificate of proof of validity of forest produce) for sawn timber, issued in the kabupaten of Tanah Laut, South Kalimantan. The certificate is valid for only five days, from 19–23 November 2002, meaning that the carrying vessel must arrive in port in Java within that time. The period of validity starts when the certificate is issued and the vessel cannot leave port before that date.

The top half of the first page states the details of the seller (PT Bahagia Setia, of Tanah Laut, South Kalimantan), the buyer of the timber (PT Kayan Jaya Tanjung, of Surabaya), and the means of transport, the vessel *Nor Buat*. The letters PL (*Perahu Layar*) before the vessel name indicate that the carrying vessel is a sailing vessel, supposedly with no engine (if it had an engine the prefixing letters should be KLM, for *Kapal Layar Motor*). However, the *Nor Buat* does have an engine as its primary means of propulsion. Also on the first page, in the box, the species and the amount of timber are stated: 500 pieces of *terantang* with a volume of 18.0864 cubic metres and 1,999 pieces of *mahang* with a volume of 11.994 cubic metres.[1] The total volume of the consignment is thus 30.0804 cubic metres. In fact, the capacity of the *Nor Buat* is more than twice this, so the greater part of the actual cargo would have been undeclared and therefore illegal. The right-hand column of the box states that the PSDH (forest resources levy) and the DR (reforestation fund) fees have been paid, but not the amounts. Unless these fees have been paid, the certificate will not be issued.

The second page provides a detailed description of the consignment. For example, for terantang, it is stated that there are 127 pieces 4.0 metres long by 20 cm wide and 6 cm thick, 130 pieces 4.0 metres long by 12 cm wide by 8 cm thick, and so on. The entire consignment should agree with the details on this page. A record of the SKSHH will be kept in the Forestry Office in the destination port and any onward movement (by truck) of this shipment or part thereof must also be congruent with the details stated on the permit.

---

[1] Terantang is possibly *dipterocarpus caudiferus*, a kind of keruing. Mahang (*macaranga* sp.) is sometimes classed as a kind of kapur.

Appendix 3

The 'small ship' pass and gross tonnage

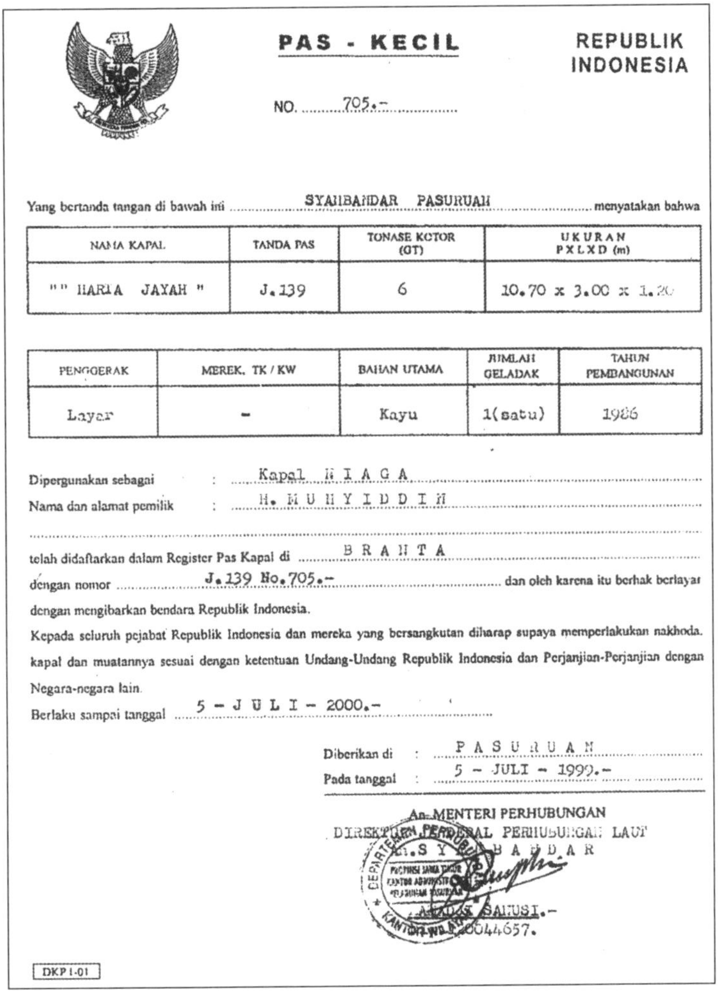

PAS - KECIL

REPUBLIK INDONESIA

NO. 705.-

Yang bertanda tangan di bawah ini SYAHBANDAR PASURUAH menyatakan bahwa

NAMA KAPAL	TANDA PAS	TONASE KOTOR (GT)	UKURAN PXLXD (m)
"" HARIA JAYAH "	J.139	6	10.70 x 3.00 x 1.20

PENGGERAK	MEREK. TK / KW	BAHAN UTAMA	JUMLAH GELADAK	TAHUN PEMBANGUNAN
Layar	-	Kayu	1(satu)	1986

Dipergunakan sebagai : Kapal N I A G A

Nama dan alamat pemilik : H. M U H Y I D D I N

telah didaftarkan dalam Register Pas Kapal di B R A N T A

dengan nomor J.139 No.705.- dan oleh karena itu berhak berlayar dengan mengibarkan bendara Republik Indonesia.

Kepada seluruh pejabat Republik Indonesia dan mereka yang bersangkutan diharap supaya memperlakukan nakhoda, kapal dan muatannya sesuai dengan ketentuan Undang-Undang Republik Indonesia dan Perjanjian-Perjanjian dengan Negara-negara lain.

Berlaku sampai tanggal 5 - J U L I - 2000.-

Diberikan di : P A S U R U A N

Pada tanggal : 5 - JULI - 1999.-

An. MENTERI PERHUBUNGAN

DIREKTUR JENDERAL PERHUBUNGAN LAUT

S Y A H B A N D A R

SAMUSI.-

DKP I-01

The *Pas Kecil*, or small ship pass, is the operating permit for small vessels. The above example is from a Madurese vessel well known to the author. According to this document, the vessel concerned, *Harta Jaya*, is 10.7 metres in length, 3.0 metres in beam and 1.2 metres in hold depth, with a capacity of six gross tons (GT). The method of propulsion (the box "Penggerak") is shown as by sail (*layar*), while the next box, "Merek, TK/KW", referring to the make and power of an engine if fitted, is marked as blank. The construction material of the vessel is timber, it has a single-level deck, and it was built in 1986. The port of registration is Branta, Madura.

The vessel is in fact considerably larger than indicated. A vessel of six gross tons should have a maximum capacity of about eighteen cubic metres of timber, but *Harta Jaya* carries more than six times that amount. The actual dimensions are approximately twenty-two metres long (excluding the bowsprit), seven metres wide, and a hold depth of well over two metres. There are sails but they are rarely hoisted, and the main power comes from a six-cylinder diesel engine. The vessel was built not in 1986, but in 1998–99 (and was observed by the author during construction). The owner had not previously owned a vessel with the same name.

This document is not the same as the vessel's measurement certificate (*surat ukur*), for which actual measurements are taken by an official. The reason for the gross understatement of the size on the small ship pass is to enable the owner to pay a minimal rate for using a port. In West Java and Central Java this document would not be accepted as proof of a capacity of only six gross tons, and the vessel's measurement certificate would be demanded. However, in Pasuruan, where this pass was issued, the stated tonnage was still being accepted in 2003. The fictitious date of construction is presumably intended to fit with the fiction concerning both the capacity of the vessel and its lack of an engine.

Such understatement of vessel size on passes was very common in Pasuruan. It was not corruption in the usual sense, because the owner was emphatic that he did not request anything untoward and that he paid only the standard amount. Rather, this is an example of a fairly widespread culture of sympathy within the Department of Marine Communications towards Pelayaran Rakyat vessel operators, many of whom operate under difficult economic conditions. Such understatement of vessel capacity does have significant statistical implications, however. Official statistics on timber discharged in ports are based on vessel manifests, which in turn reflect the volumes shown on SKSHH (timber transport permit) documents. Since most manifests show volumes well short of the actual volumes carried, the statistics are unrealistic. For the researcher, the way around this incongruity is to rely not on official data for timber discharged, but rather on the gross tonnage of vessels. The rough rule of thumb used by harbourmaster's office staff for their own purposes is to multiply the registered gross tonnage by three, to get the volume of timber carried in cubic metres. In practice, timber does not stow that well, with a stowage factor of about two cubic metres to the gross ton (2.83 cubic metres of cargo space), but this error is more or less cancelled out by the large amount of timber carried on the deck. Unfortunately, this approach is also likely to give a flawed result because the gross tonnages themselves are often inaccurate.

Not all vessels using Pasuruan have understated gross tonnages. In the case of vessels from Sumenep, for example, the tonnages are generally accurate. On the other hand, even in Jakarta large understatements of gross tonnage are not uncommon. A very large and nearly new two-masted Bugis vessel observed in 2003 had a reported — and believable — capacity of 1,500 cubic metres but a registered gross tonnage of only 172! According to the manager of the Pelra office at Sunda Kelapa, such cases are 'manipulated' to reduce port costs. While such understated gross tonnages were also formerly convenient to conceal the actual amount of timber carried, these days Airud water police and port forestry officers are much more aware of the real capacity of vessels.

Bibliography

Abdurachman (1971), *Sejarah Madura Selayang Pandang* (5th edn). Sumenep: Publisher n.k.

——— (1979), "Masalah Carok di Madura", in Samsuri (ed.), *Madura II*. Malang: Proyek Penelitian Madura, Departemen Pendidikan dan Kebudayaan, IKIP Malang.

Aditjondro, George J. (1998), *Mengenal Penjarah Kekayaan Negara*. Jakarta: Kelompok Studi Reformasi.

Amarell, G. (1999), *Bugis Navigation*. Yale University Southeast Asian Studies, monograph no. 48. New Haven: Yale University Press.

Aspinall, E. and G. van Klinken (2011), *The State and Illegality in Indonesia*. Leiden: KITLV Press.

BPS Bangkalan (2001), *Bangkalan dalam Angka 2001*. Bangkalan: Badan Pusat Statistik.

BPS Pamekasan (2000), *Pamekasan dalam Angka 2000*. Pamekasan: Badan Pusat Statistik.

BPS Sampang (2000), *Sampang dalam Angka 2000*. Sampang: Badan Pusat Statistik.

BPS Sumenep (2000), *Sumenep dalam Angka 2000*. Sumenep: Badan Pusat Statistik.

Baradas, David B. (1968), "Some Implications of the Okir Motif in Lanao and Sulu Art". *Asian Studies*, VI, no. 1: 129–68.

Barnes, R.H. (1993), "Salt Production in East Flores Regency, Nusa Tenggara Timur, Indonesia", in P. LeRoux and J. Ivanoff (eds), *Le sel de la vie en Asie Sud-est*. Bangkok: Prince of Songkla University/White Lotus Books.

Barr, C. and I.A.P. Resosudarmo (2001), "Overview of Indonesia's Decentralisation Process". Preface to A. Casson, *Decentralisation of Policies Affecting Forests and Estate Crops in Kotawaringin Timur District, Central Kalimantan*. Bogor: Center for International Forestry Research.

Blake, W. (1925), "The Madura Prao". *Yachting*, January 1925: 55, 98.

Boomgaard, P. (1988), "Forests and Forestry in Colonial Java: 1677–1942", in J. Dargavel, K. Nixon and N. Semple (eds), *Changing Tropical Forests: Historical Perspectives on Today's Challenges in Asia, Australasia and Oceania*. Canberra: Centre for Resource and Environment Studies, Australian National University.

_______ (1991), "The Non-agricultural Side of an Agricultural Java, 1500–1900", in P. Alexander, P. Boomgaard and B. White (eds), *In the Shadow of Agriculture: Non-farm Activities in the Javanese Economy, Past and Present*. Amsterdam: Royal Tropical Institute.

_______ (1996), "Forests and Forestry 1823–1941". *Changing Economy in Indonesia*, vol. 16. Amsterdam: Royal Tropical Institute.

Braddon, Russell (1986), *Thomas Baines and the North Australian Expedition*. Sydney: Collins, in association with the Royal Geographical Society, London.

Brissenden, Rosemary (1976), "Patterns of Trade and Maritime Society before the Coming of the Europeans", in Elaine McKay (ed.), *Studies in Indonesian History*. Melbourne: Pitman.

Bruyn Kops, G.F. de (1854), "Iets over de Zeevaart in den Indischen Archipel". *Tijdschrift voor Nijverheid in Nederlandsch-Indië*, part 1: 21–69, 97–135.

Burningham, N. and J. Mellefont (1996), "Indonesian Seacraft: Photographs and Notes by G.C. Ingleton, 1932". *The Great Circle*, 18 (2): 97–112.

_______ (1997), "The Exceptional *Janggolan*: Engineless Sailing Ships Still Trading from Madura, Indonesia". *Bulletin of the Australian Institute for Maritime Archaeology*, 21 (1 & 2): 35–66.

Burningham, N. and K. Stenross (1994), "*Mayang*: The Traditional Fishing Vessel of Java". *The Beagle: Records of the Museums and Art Galleries of the Northern Territory*, 11: 73–132.

Butcher, J. and H.W. Dick (eds) (1993), *The Rise and Fall of Revenue Farming: Business Elites and the Emergence of the Modern State in Southeast Asia*. New York: St Martin's Press.

Campo, J.N.F. à (1993), "Perahu shipping in Indonesia 1870–1914". *Review of Indonesian and Malaysian Affairs*, 27 (1 and 2): 33–60.

Casson, A. (2001), *Decentralisation of Policies Affecting Forests and Estate Crops in Kotawaringin Timur District, Central Kalimantan*. Decentralisation and Forests in Indonesia series, case study no. 5. Bogor: Center for International Forestry Research.

Castles, L. (1967), *Religion, Politics and Economic Behaviour in Java: The Kudus Cigarette Industry*. New Haven: Yale University Southeast Asian Studies.

Cense, A.A. (1972), *Pelayaran dan Pengaruh Kebudayaan Makassar-Bugis di Pantai Utara Australia*. Jakarta: Bharata.

Chudnoff, Martin (1980), *Tropical Timbers of the World*. Madison: Forest Products Laboratory, U.S. Department of Agriculture.

Collins, G.E.P. (1936), *East Monsoon*. London: Jonathan Cape.

_______ (1992), *Makassar Sailing*. Singapore: Oxford University Press [First published in 1937 by Jonathan Cape, London].

Collins, N.M., J.A. Sayer and T.C. Whitmore (eds) (1991), *The Conservation Atlas of Tropical Forests: Asia and the Pacific*. London: Macmillan.

Cordes, Th. (1937), "Visscherij, Veeteelt en Vrachtvaart op Madoera". *Blaadje voor het Volkscredietwezen*, 25: 683–96.

Cortesao, A. (ed., transl.) (1967), *The Suma Oriental of Tomé Pires*. Liechtenstein: Kraus Reprint [First published in 1944 by the Hakluyt Society, London].

Cribb, Robert (2000), *Historical Atlas of Indonesia*. Richmond, UK: Curzon Press.

Dauvergne, Peter (1997), *Shadows in the Forest: Japan and the Politics of Timber in Southeast Asia*. Cambridge, Mass.: MIT Press.

Davies, William (1999), *Madurese*. Munchen: Lincom Europa.

Deventer, C. Th. van (1904), *Overzicht van den Economischen Toestand der Inlandsche Bevolking van Java en Madoera*. s'Gravenhage: Martinus Nijhoff.

Dick, H.W. (1975a), "Prahu Shipping in Eastern Indonesia", Part I. *Bulletin of Indonesian Economic Studies*, 11 (2): 69–107.

——— (1975b), "Prahu Shipping in Eastern Indonesia", Part II. *Bulletin of Indonesian Economic Studies*, 11 (3): 81–103.

——— (1978), "Indonesian Prahus and Chinese Junks: Some Issues of Definition and Policy in Informal Sector Studies", in P.J. Rimmer, D.W. Drakakis-Smith and T.G. McGee (eds), *Food, Shelter and Transport in Southeast Asia and the Pacific*. Canberra: Research School of Pacific Studies, Australian National University.

——— (1980), "The Rise and Fall of Dualism", in R.G. Garnaut and P.T. McCawley (eds), *Indonesia: Dualism, Growth and Poverty*. Canberra: Research School of Pacific Studies, Australian National University.

——— (1987), "Prahu Shipping in Eastern Indonesia in the Interwar Period". *Bulletin of Indonesian Economic Studies*, 23 (1): 104–21.

——— (1993), "Indonesian Economic History Inside Out". *Review of Indonesian and Malaysian Affairs*, 27 (1 and 2): 1–12.

——— (2002), *Surabaya, City of Work: A Socioeconomic History, 1900–2000*. Athens, Ohio: Ohio University Press.

——— (2008), "The 2008 Shipping Law: De-regulation or Re-regulation?" *Bulletin of Indonesian Economic Studies*, 44 (3).

Dick, H.W. and P.J. Rimmer (2003), *Cities, Transport and Communications: The Integration of Southeast Asia since 1850*. Basingstoke: Palgrave Macmillan.

EIA (Environmental Investigation Agency) (1998), *The Politics of Extinction*. Report, at http://www.eia-international.org./Campaigns/Forests/Indonesia/PolExtinction/index.html, 02/04/2002.

——— (2001), *The Final Cut: Illegal logging in Indonesia's Orang-utan Parks*. Report, at http://www.eia-international.org/Campaigns/Forests/Indonesia/FinalCut/index.html, 17/10/2001.

Earl, G.W. (1971), *The Eastern Seas*. New York and Singapore: Oxford University Press [First published in 1837 by W.H. Allen, London].

Elson, R.E. (1984), *Javanese Peasants and the Colonial Sugar Industry: Impact and Change in an East Java Residency, 1830–1940*. Singapore: Oxford University Press.

Emmerson, Donald K. (1980), "The Case for a Maritime Perspective on Southeast Asia". *Journal of Southeast Asian Studies*, XI (1): 139–45.

ENI (*Encyclopaedie van Nederlandsche-Indië*) (8 volumes, 1917–1927). s'Gravenhage and Leiden: Martinus Nijhoff and E.J. Brill.

Ensiklopedi Indonesia (1983). Jakarta: Buku Ichtiar Baru/van Hoeve.

Evers, Hans-Dieter (1983), "Traditional Trading Networks of Southeast Asia". *Archipel*, 35: 89–100.

Feldman, Jerome (ed.) (1985), *The Eloquent Dead: Ancestral Sculpture of Indonesia and Southeast Asia*. Los Angeles: UCLA Museum of Natural History.

Feretti, Eveline (ed.) (1997), *Cutting Across the Lands: An Annotated Bibliography on Natural Resource Management and Community Development in Indonesia, the Philippines and Malaysia*. Ithaca, NY: Southeast Asia Program Publications, Cornell University Press.

Geertz, C. (1963), *Peddlers and Princes: Social development and Economic Change in Two Indonesian Towns*. Chicago: University of Chicago Press.

Gibson-Hill. C.A. (1949), "Cargo Boats of the East Coast of Malaya". *Journal of the Malayan Branch of the Royal Asiatic Society*, 22 (3): 106–25.

________ (1950), "The Indonesian Trading Boats Reaching Singapore". *Journal of the Malayan Branch of the Royal Asiatic Society*, 23 (1): 108–38.

________ (1952), "Tongkang and Lighter Matters". *Journal of the Malayan Branch of the Royal Asiatic Society*, 25 (1): 84–110.

Gillis, Malcolm (1988), "Indonesia: Public Policies, Resource Management, and the Tropical Forest", in R. Repetto and M. Gillis (eds), *Public Policies and the Misuse of Forest Resources*. New York: Cambridge University Press.

Hageman, J. Jcz (1858), "Bijdragen tot de Kennis van de Residentie Madoera". *Tijdschrift voor Nederlandsch-Indië*, 20 (1): 321–52.

________ (1862), *Over de Eilanden van Madoera*. Leiden: MS KITLV, D H-16.

________ (1863), "Aantekeningen over Nijverheid en Landbouw op Madoera". *Tijdschrift voor Nijverheid en Landbouw in Nederlandsch-Indië*, 9: 272–95.

Hawkins, Clifford W. (1982), *Praus of Indonesia*. London: Nautical Books.

Hornell, J. (1920), "The Outrigger Canoes of Indonesia". Report no. 2, *Madras Fisheries Bulletin*, vol. XII: 43–114.

________ (1970), *Water Transport: Origins and Early Evolution*. Newton Abbot, UK: David and Charles [First published in 1946 by Cambridge University Press, Cambridge, and Bentley House, London].

Horridge, G.A. (1978), *The Design of Planked Boats of the Moluccas*. Maritime Monographs and Reports, no. 38. London: National Maritime Museum.

________ (1979a), *The Lambo or Prahu Bot: A Western Ship in an Eastern Setting*. Maritime Monographs and Reports, no. 39. London: National Maritime Museum.

———— (1979b), *The Konjo Boatbuilders and the Bugis Prahus of South Sulawesi*. Maritime Monographs and Reports, no. 40. London: National Maritime Museum.

———— (1981), *The Prahu: Traditional Sailing Boat of Indonesia*. Kuala Lumpur: Oxford University Press.

———— (1982), *The Lashed-lug Boat of the Eastern Archipelagoes*. Maritime Monographs and Reports, no. 54. London: National Maritime Museum.

———— (1986), *Sailing Craft of Indonesia*. Singapore: Oxford University Press.

Hughes, David E. (1984), 'The Indonesian Cargo Sailing Vessels and the Problem of Technology Choice for Sea Transport in a Developing Country: A Study in the Consequences of Perahu Motorisation Policy in the Context of the Economic Regulation in Inter-island Shipping'. Unpublished Ph.D. dissertation, University of Wales, UK.

———— (1986), "The Prahu and Unrecorded Inter-island Trade". *Bulletin of Indonesian Economic Studies*, 22 (2): 103–13.

Huitema, T. (1995), *Ronde en platbodemjachten*. Baam, Neth.: Uitgeverij Hollandia.

Husson, Laurence (1991), "Madura et le *Carok*. Violence Coutumière et Monde Moderne". *Archipel*, 40: 29–39.

———— (1997), "Eight Centuries of Madurese Migration to East Java". *Asia and Pacific Migration Journal*, 6 (1): 77–102.

Indonesia Pilot (1976), vol. II. London: Hydrographer of the Navy.

Jonge, Huub de (1989), *Madura dalam Empat Zaman: Pedagang, Perkembangan Ekonomi dan Islam*. Seri terjemahan KITLV/LIPI. Jakarta: Gramedia.

———— (1993), "Monopolization and Resistance: State and Salt Producers in Madura", in P. Le Roux and J. Ivanoff (eds), *Le Sel de la Vie en Asie du Sud-est*. Bangkok: Prince of Songkla University/White Lotus Books.

———— (1995), "Stereotypes of the Madurese", in K. van Dijk, H. de Jonge and E. Touwen-Bouwsma (eds), *Across Madura Strait: The Dynamics of an Insular society*. Leiden: KITLV Press.

Jonge, Huub de and Gerben Nooteboom (2006), "Why the Madurese? Ethnic Conflict in West and East Kalimantan Compared". *Asian Journal of Social Science*, 34 (3): 456–74.

Jordaan, R. and A. Niehof (1980), "Aspects of Fishing in Patondu, a Village on the North Coast of Madura". *Review of Indonesian and Malaysian Affairs*, 14 (1): 81–111.

———— (1982), "Patondu Revisited: A Case Study of Modernization in Fishery". *Review of Indonesian and Malaysian Affairs*, 16 (2): 83–108.

Kehutanan (1974), *Lampiran-lampiran: Bahan-bahan Rapat Kerja Evaluasi Pemasukan Kayu ke Pulau Jawa dan Bali*. Compilation of papers presented at a Workshop on the Importation of Timber into Java and Bali, held in Yogyakarta, 26–27 August 1974. Jakarta: Direktorat-Jenderal Kehutanan.

A: "Pidato Pengarahan Direktur Jenderal Kehutanan". Direktur-Jenderal Kehutanan.

B: "Kondisi dan Situasi Pemasukan Kayu ke P. Jawa, Madura dan Bali". Laporan, Direktorat Pemasaran Kehutanan.

C: "Perdagangan Hasil Hutan dalam Propinisi Kalimantan Selatan". Laporan, Dinas Kehutanan Kalimantan Selatan.

D: "Perdagangan Kayu Rimba Asal Luar Jawa dan Peranan Perum Perhutani dalam Distribusi di Jawa". Laporan, Direksi Perum Perhutani.

E: "Rapat Kerja Evaluasi Pemasukan Kayu ke Jawa dan Bali". Laporan, Dinas Kehutanan DKI Jakarta Raya.

F: "Unit I Perum Perhutani Jawa Tengah". Laporan, Unit I Perum Perhutani Jawa Tengah.

Kemp, P.H. van der (1894), *Handboek tot de Kennis van's Lands Zoutmiddel in Nederlandsch-Indië*. s'Gravenhage: Kolff and Co.

Kiefer, Thomas M. and Clifford Sather (1970), "Gravemarkers and the Repression of Sexual Symbolism: The Case of Two Philippine-Borneo Moslem Societies". *Bijdragen*, 126: 75–90.

Klinken, G. van (2002), "Indonesia's New Ethnic Elites", in H. Schulte Nordholt and Irwan Abdullah (eds), *Indonesia in search of transition*. Yogyakarta: Pustaka Pelajar.

Knaap, G. (1996), *Shallow Waters, Rising Tide: Shipping and Trade in Java around 1775*. Leiden: KITLV Press.

Knaap, G. and L. Nagtegaal (1991), "A Forgotten Trade: Salt in Southeast Asia 1670–1813", in R. Ptak and D. Rothermund (eds), *Emporia, Commodities and Entrepreneurs in Asian Maritime Trade, c. 1400–1750*. Stuttgart: Franz Steiner Verlag.

Knaap, G. and H. Sutherland (2004), *Monsoon Traders: Ships, Skippers and Commodities in Eighteenth Century Makassar*. Leiden: KITLV Press.

Koesnadi (2002), *Konflik Sosial Nelayan*. Yogyakarta: LKIS.

Kuntowijoyo (1980), *Social Change in an Agrarian Society: Madura, 1850–1940*. Ph.D. dissertation, Columbia University. Ann Arbor: UMI.

Kurlansky, M. (2002), *Salt: A World History*. London: Jonathan Cape.

Lapian, A.B. (1979), "Le rôle des Orang Laut dans l'histoire de Riau". *Archipel*, 18: 215–22.

Latief Wiyata, A. (2002), *Carok. Konflik Kekerasan dan Harga Diri Orang Madura*. Yogyakarta: LKIS.

Leirissa, R.Z. (1993), "The Structure of Makassar-Bugis Trade in the Pre-modern Moluccas". *Review of Indonesian and Malaysian Affairs*, 27 (1 and 2): 77–90.

Lindblad, J. Thomas (1988), *Between Dayak and Dutch: The Economic History of Southeast Kalimantan*. Dordrecht: Foris.

Lips, M.E.M. (1993), 'Van Perahu naar Kapal'. Unpublished Ph.D. dissertation, Twente University, Netherlands.

Lopa, Baharuddin (1982), "Hukum Laut, Pelayaran dan Perniagaan: Penggalian dari Bumi Indonesia Sendiri". Unpublished Ph.D. dissertation, Diponegoro University, Semarang.

Macknight, C. (1976), *The Voyage to Marege: Macassan Trepangers in Northern Australia*. Melbourne: Melbourne University Press.

——— (1980), "The Study of Praus in the Indonesian Archipelago". *The Great Circle*, 2 (2): 117–28.

Macknight, C. and Mukhlis (1979), "A Bugis Manuscript about Praus". *Archipel*, 18: 225–82.

Magenda, Burhan (1991), *East Kalimantan: The Decline of a Commercial Aristocracy*. Ithaca, NY: Cornell University Press.

Mansurnoor, Iik Arifin (1990), *Islam in an Indonesian World: Ulama of Madura*. Yogyakarta: Gajah Mada University Press.

——— (1995), "Rato and Kiai in Madura: Are they twins?" in K. van Dijk, H. de Jonge and E. Touwen-Bouwsma (eds), *Across Madura Strait: The Dynamics of an Insular Society*. Leiden: KITLV Press.

Mansvelt, W.M.F. (1938), "De Prauwvaart in de 19de Eeuw". *Koloniale Studiën*, 22: 89–102.

Masyhuri (1996), *Menyisir Pantai Utara: Usaha dan Perekonomian Nelayan di Jawa dan Madura 1850–1940*. Yogyakarta: Yayasan Pustaka Nusantara with KITLV, Leiden.

McCarthy, J.F. (2001a), *Decentralisation, Local Communities and Forest Management in Barito Selatan District, Central Kalimantan*. Decentralisation and Forests in Indonesia series, case study no. 1. Bogor: Center for International Forestry Research.

——— (2001b), *Decentralisation and Forest Management in Kapuas District, Central Kalimantan*. Decentralisation and Forests in Indonesia series, case study no. 2. Bogor: Center for International Forestry Research.

Meilink-Roelofsz, M.A.P. (1962), *Asian Trade and European Influence in the Indonesian Archipelago between 1500 and about 1630*. The Hague: Martinus Nijhoff.

Mellefont, J. (1991), "Between Tradition and Change: An Indonesian Perahu in an Australian Collection". *Great Circle*, 13 (2): 97–110.

Mills, J.V. (ed.) (1930), "Eredia's Description of Malacca, Meridional India, and Cathay". *Journal of the Malayan Branch of the Royal Asiatic Society*, 8 (1): 1–288.

Muthmainnah (1998), *Jembatan Suramadu: Respon Ulama terhadap Industrialisasi*. Jakarta: LKPSM.

Nagtegaal, Luc (1996), *Riding the Dutch Tiger: The Dutch East Indies Company and the Northeast Coast of Java, 1680–1743*. Leiden: KITLV Press.

Nieuwenkamp, W.O.J. (1926), "Vaartuigen in Tropisch Nederland". *Nederlandsche-Indië Oud en Nieuw*, 11: 105–11, 145–55, 172–77, 208–12.

Nooteboom, C. (1940), "Vaartuigen van Mandar". *Tijdschrift voor Indische Taal-, Lande-, en Volkenkunde*, 80: 22–33.

Nooteboom, Gerben (2008), "Through Turbulent Times: Diversity, Vulnerability and Resilience of Madurese Livelihoods in East Kalimantan", in M. Titus and P. Burgers (eds), *Rural Livelihoods, Resources and Coping with Crisis in Indonesia*. Singapore: ISEAS/Amsterdam University Press.

_______ (2010), "Out of Wedlock: Migrant-Police Partnerships in East Kalimantan", in E. Aspinall and G. van Klinken (eds), *State and Illegality in Indonesia*. Leiden: KITLV Press.

Nordholt, Henk Schulte (1981), "The Mads Lange Connection. A Danish Trader in Bali in the Middle of the Nineteenth Century: Broker and Buffer". *Indonesia*, 32: 16–47.

_______ (2000), "A state of violence" (excerpts from oration text). IASS Newsletter, 23.

OMW (Onderzoek naar de Mindere Welvaart der Inlandsche Bevolking op Java en Madoera) (1905), *Samentrekking der Afdeelingsverslagen over de Uitkomsten der Onderzoekingen naar de Visschteelt en Visscherij in de Residentie Madoera*. Batavia: Landsdrukkerij.

_______ (1906), *Samentrekking der Afsdeelingsverslagen over de Uitkomsten der Onderzoekingen naar het Vervoerwezen in de Residentie Madoera*. Batavia: F.B. Smits.

_______ (1907), *Overzicht van de Uitkomsten der Gewestelijke Onderzoekingen naar het Vervoerwezen en Daaruit Gemaakte Gevolgtrekkingen*. Batavia: G. Kolff and Co.

_______ (1909), *Overzicht van de Uitkomsten der Gewestelijke Onderzoekingen naar den Inlandschen Handel en Nijverheid en Daaruit Gemaakte Gevolgtrekkingen* (deel 2, bijlagen). Batavia: H.M. van Dorp.

Ormeling, F.J. (1956), *The Timor Problem: A Geographic Interpretation of an Undeveloped Island*. Groningen: Wolters.

Paris, M. (1842), *Essaie sur la Construction Navale des Peuples Extra-Européens*. Paris: Arthus Bertrand.

Pelly, Usman (1975), *Ara dengan Perahu Bugisnya*. Ujung Pandang: Pusat Latihan Penelitian Ilmu-ilmu Sosial, Universitas Hasanuddin.

_______ (1977), "Symbolic Aspects of the Bugis Ship and Shipbuilding". *Journal of the Steward Anthropological Society*, 8: 87–106.

Pelras, Christian (1996), *The Bugis*. Oxford: Blackwell.

Peluso, Nancy Lee (1983), *Markets and Merchants: The Forest Products Trade of East Kalimantan in Historical Perspective*. M.A. dissertation, Cornell University. Ann Arbor: UMI.

_______ (1988), *Rich Forests, Poor People and Development: Forest Access, Control and Resistance in Java*. Ph.D. dissertation, Cornell University. Ann Arbor: UMI.

_______ (1990), "A History of State Forest Management in Java", in M. Poffenberger (ed.), *Keepers of the Forest: Land Management Alternatives in Southeast Asia*. West Hartford, CT: Kumarian Press.

Perla (2002), *Persyaratan Keselamatan bagi Kapal Layar Motor (KLM) berukuran Tonase Kotor sampai dengan GT 500*. Jakarta: Direktorat-Jenderal Perhubungan Laut.

Petebang, E. and E. Sutrisno (2000), *Konflik Etnik di Sambas*. Jakarta: Institute Studi Arus Informasi.

Piollet, Paul (1995), *Equipages et Voiliers de Madura*. Ternant: Piollet.

Potter, Lesley (1988), "Indigenes and Colonisers: Dutch Forest Policy in South and East Borneo", in J. Dargavel, K. Dixon and N. Semple (eds), *Changing Tropical Forests: Historical Perspectives on Today's Challenges in Asia, Australasia and Oceania*. Canberra: Centre for Resource and Environmental Studies, Australian National University.

——— (1991), "Environmental and Other Social Aspects of Timber Exploitation in Kalimantan, 1967–1989", in Joan Hardjono (ed.), *Indonesia: Resources, Ecology and Environment*. New York: Oxford University Press.

Rachbini, Didik J. (1995), "Conditions and Consequences of Industrialization in Madura", in K. van Dijk, H. de Jonge and E. Touwen-Bouwsma (eds), *Across Madura Strait: The Dynamics of an Insular Society*. Leiden: KITLV Press.

Raffles, T.S. (1978), *The History of Java*, vol. I. Kuala Lumpur: Oxford University Press [First published in 1817 by Black, Parbury and Allen, London].

Regeeringsalmanak voor Nederlandsch-Indië (1865–1942). Batavia: Landsdrukkerij.

Reid, A. (1993), "Islamization and Christianization in Southeast Asia: The Critical Phase, 1550–1650", in A. Reid (ed.), *Southeast Asia in the Early Modern Era: Trade, Power, and Belief*. Ithaca, NY: Cornell University Press.

Ricklefs, M. (1993), *A History of Modern Indonesia* (2nd edn). London: Macmillan.

Robison, R. (1986), *Indonesia: The Rise of Capitalism*. Sydney: Allen and Unwin.

Roche, Patrick A. (1984), *Fishermen of the Coromandel: A Social Study of the Paravas of Coromandel*. New Delhi: Manohar.

Ross, Michael L. (2001), *Timber Booms and Institutional Breakdown in Southeast Asia*. New York: Cambridge University Press.

Rouffaer, G.P. (1904), "De Voornaamste Industrieën der Inlandsche Bevolking van Java en Madoera". Published as appendix to C. Th. van Deventer, *Overzicht van den Economischen Toestand der Inlandsche Bevolking van Java en Madoera*. s'Gravenhage: Martinus Nijhoff.

Schrieke, B. (1957), *Indonesian Sociological Studies, Part II: Ruler and Realm in Early Java*. The Hague: van Hoeve.

——— (1966), *Indonesian Sociological Studies, Part I*. The Hague: van Hoeve.

Scott, W.H. (1982), "Plank-built boats in the Philippines". *Philippine Studies*, 30: 335–76.

Smith, Glenn (1989), "Pentingnya Sapi dalam Masyarakat Madura", in H. de Jonge (ed.), *Agama, Kebudayaan dan Ekonomi: Studi-studi Interdisipliner tentang Masyarakat Madura*. Jakarta: Rajawali Pers.

_______ (1995), "Madurese Maize and Bovines Seen through an Ecological-materialist Lens", in K. van Dijk, H. de Jonge and E. Touwen-Bouwsma (eds), *Across Madura Strait: The Dynamics of an Insular Society*. Leiden: KITLV Press.

_______ (ed.) (2001), *Bibliography of Madura (Indonesia)*. At http://coombs.anu.edu.au/Biblio/biblio_madura2.html, 20/03/2002.

Soewanda, R. Among Prawira *et al.* (1971), *Daftar Nama Pohon-pohonan, Samarinda (Kalimantan Timur)*. Bogor: Lembaga Penelitian Hutan.

Southon, M. (1995), *The Navel of the Perahu: Meaning and Values in the Maritime Trading Economy of a Butonese Village*. Canberra: Research School of Pacific and Asian Studies, Australian National University.

Steinberg, D.J. (ed.) (1987), *In Search of Southeast Asia: A Modern History* (revised edn). St Leonards, NSW: Allen and Unwin/University of Hawai'i Press.

Sudagung, Hendro Suroyo (2001), *Menguraikan Pertikaian Etnis: Swakarsa Etnis Madura ke Kalimantan Barat*. Jakarta (?): Institut Studi Arus Informasi, in association with Yayasan Adikarya IKAPI and the Ford Foundation.

Sularto, Hadisuwarno (1985), *Penelitian Tentang Konstruksi Perahu Layar Tradisional*. Ujung Pandang: Lembaga Penelitian, Hasanuddin University.

Sulaiman, B.A. (1980), *Seni Ukir Madura*. Jakarta: Direktorat-Jenderal Kebudayaan.

Surata, Agus and Tuhana Taufiq Andrianto (2001), *Atasi Konflik Etnis*. Yogyakarta: Global Pustaka Utama.

Surink, H.A (1926), "De Vischvangst op Madoera". *Djawa*, VI: 266–70.

_______ (1932), "Het Zout van Madoera". *Tijdschrift voor Economische Geographie*, 23: 250–54.

Tagliacozzo, E. (2005), *Secret Trades, Porous Borders: Smuggling and States along a Southeast Asian Frontier, 1865–1915*. New Haven: Yale University Press. Also (2007), Singapore: NUS Press.

Thomasz, L.F.R. (1993), "The Malay Sultanate of Melaka", in A. Reid (ed.), *Southeast Asia in the Early Modern Era: Trade, Power, and Belief*. Ithaca, NY: Cornell University Press.

Tobing, Ph. L. (1961), *Hukum Pelayaran Amanna Gappa*. Ujung Pandang: Yayasan Kebudayaan Sulawesi Selatan dan Tenggara.

Touwen-Bouwsma, E. (1988), *Staat, Islam en Locale Leiders in West Madura, Indonesie*. Ph.D. dissertation, Amsterdam: Uitgeverij Mondiss.

_______ (1989), "Kepala Desa Madura: Dari Boneka ke Wiraswasta", in H. de Jonge (ed.), *Agama, Kebudayaan dan Ekonomi: Studi-studi Interdisipliner tentang Masyarakat Madura*. Jakarta: Rajawali Pers.

Verwijk, J.J. (1893), "Bijdrage tot de Kennis van den Sapoedi-archipel". *Tijdschrift voor Indische Taal-, Lande-, en Volkenkunde*, 35: 234–38.

Vuuren, L. van (1916–1917), "De Prauwvaart van Celebes". *Kolonial Studiën*, 1: 107–16; 2: 329–39.

Warren, J.F. (1981), *The Sulu Zone 1768–1898: The Dynamics of External Trade, Slavery and Ethnicity in the Transformation of a Southeast Asian Maritime State*. Singapore: Singapore University Press.

Film

The Hasans: A Buginese Trading Family. Film Australia, 1974.

Newspapers and periodicals (cited without reference to author)

Bali Post
Borneo (PT Mentaya Grafika Press, Sampit)
Forum Keadilan
Jakarta Post
Tempo

Index